TEXTBOOK OF AGRONOMY

TEXTBOOK OF AGRONOMY

GYAN DEEP SINGH

ANMOL PUBLICATIONS PVT. LTD.
NEW DELHI - 110 002 (INDIA)

ANMOL PUBLICATIONS PVT. LTD.
H.O.: 4374/4B, Ansari Road, Daryaganj,
New Delhi-110 002 (India)
Ph.: 23278000, 23261597
B.O.: No. 1015, Ist Main Road, BSK IIIrd Stage
IIIrd Phase, IIIrd Block,
Bangalore - 560 085 (India)
Visit us at: www.anmolpublications.com

Textbook of Agronomy

ISBN 978-81-261-3598-1

PRINTED IN INDIA

Printed at Mehra Offset Press, Delhi.

Contents

Preface

In this period of competitiveness, one has to be updated with the latest knowledge. Hence in this second edition, efforts have been made to update the available information. The first chapter, field crops-overview, has been completely revised incorporating the latest available data. Additional information has been provided in the chapters such as nutrient management mainly organic farming, watershed management, farming system. Information technology in the field of agronomy has been added in the last chapter.

We hope that this book will serve as a textbook for the graduate students and teachers in agricultural institutions and those who are interested in the subject.

Author

Preface

In this period of competitiveness, one has to be updated with the latest knowledge. Hence in this second edition, efforts have been made to update the available information. The first chapter, Field crops-overview, has been completely revised incorporating the latest available data. Additional information has been provided in the chapters such as nutrient management, mainly organic farming, watershed management, farming system. Information technology in the field of agronomy has been added in the last chapter.

We hope that this book will serve as a textbook for the graduate students and teachers in agricultural institutions and those who are interested in the subject.

Authors

Chapter 1

History of Agriculture

INTRODUCTION

Agriculture, art, science, and industry of managing the growth of plants and animals for human use. In a broad sense agriculture includes cultivation of the soil, growing and harvesting crops, breeding and raising livestock, dairying, and forestry. Regional and national agriculture are covered in more detail in individual continent, country, state. Modern agriculture depends heavily on engineering and technology and on the biological and physical sciences. Irrigation, drainage, conservation, and sanitary engineering each of which is important in successful farming are some of the fields requiring the specialized knowledge of agricultural engineers. Agricultural chemistry deals with other vital farming concerns, such as the application of fertilizer, insecticides, and fungicides, soil makeup, analysis of agricultural products, and nutritional needs of farm animals.

Plant breeding and genetics contribute immeasurably to farm productivity. Genetics has also made a science of livestock breeding. Hydroponics, a method of soilless gardening in which plants are grown in chemical nutrient solutions, may help meet the need for greater food production as the world's population increases. The packing, processing, and marketing of agricultural products are closely related activities also influenced by science. Methods of quick-freezing and dehydration have increased the markets for farm products.Mechanization, the outstanding characteristic of late 19th- and 20th-century agriculture, has eased much of the backbreaking toil of the farmer. More significantly, mechanization has enormously increased farm efficiency and productivity. Animals including horses, oxen, llamas, alpacas, and dogs, however, are still used to cultivate fields, harvest crops, and transport farm products to markets in many parts of the world. Airplanes and helicopters are employed in agriculture for seeding, spraying operations for insect and disease control, transporting perishable products, and fighting forest fires. Increasingly satellites are being used to monitor crop yields. Radio and television disseminate vital weather reports and other information such as market reports that concern farmers. Computers have become an essential tool for farm management.

ANCIENT AGRICULTURE

Before agriculture, people lived by hunting wild animals and gathering edible plants. When the herds were plentiful and the plants flourishing, life was good. But, when the herds migrated elsewhere, people had to follow them and often discover a whole new set of plants to supplement their diet.

Hunters eventually realized that their prey was much easier to kill if it were walled up in a box canyon. Better yet, they could capture the prey and keep it in a cave for future use. Archaeological finds show that early humans imprisoned giant ground sloth's in this way. Entrapment, however, was a temporary measure. Not thinking of the future, hungry humans gorged themselves, then, when the sloths had all been eaten, they sought out more sloths. maintaining a herd by breeding and nurturing wasn't yet practiced. This "feast or famine" lifestyle had its definite drawbacks including starvation. Fortunately, several geniuses throughout the world eventually discovered how to preserve meat by drying it, smoking it over a fire, or cooking it. some others realized that it they took the seeds of the plants they had been eating and scattered them about, they grew into new plants.Eventually, people decided that life would be a lot easier if they always had the animals with them and if edible plants or their produce were always available. Settling down seemed like a good idea.

HISTORY

The history of agriculture may be divided into five broad periods of unequal length, differing widely in date according to region: prehistoric, historic through the Roman period, feudal, scientific, and industrial. A countertrend to industrial agriculture, known as sustainable agriculture or organic farming, may represent yet another period in agricultural history.

Prehistoric Agriculture

Early farmers were, archaeologists agree, largely of Neolithic culture. Sites occupied by such people are located in southwestern Asia in what are now Iran, Iraq, Israel, Jordan, Syria, and Turkey; in southeastern Asia, in what is now Thailand; in Africa, along the Nile River in Egypt; and in Europe, along the Danube River and in Macedonia, Thrace, and Thessaly (historic regions of southeastern Europe). Early centers of agriculture have also been identified in the Huang He (Yellow River) area of China; the Indus River valley of India and Pakistan; and the Tehuacán Valley of Mexico, northwest of the Isthmus of Tehuantepec.

The dates of domesticated plants and animals vary with the regions, but most predate the 6th millennium bc, and the earliest may date from 10,000 bc. Scientists have carried out carbon-14 testing of animal and plant remains and have dated finds of domesticated sheep at 9000 bc in northern Iraq; cattle in the 6th millennium bc in northeastern Iran; goats at 8000 bc in central Iran; pigs at 8000 bc in Thailand and 7000 bc in Thessaly; onagers, or asses, at 7000 bc in Iraq; and horses around 4000 bc in central Asia. The llama and alpaca were domesticated in the Andean regions of South America by the middle of the 3rd millennium bc.

According to carbon dating, wheat and barley were domesticated in the Middle East in the 8th millennium bc; millet and rice in China and Southeast Asia by 5500 bc; and squash in Mexico about 8000 bc. Legumes found in Thessaly and Macedonia are dated as early as 6000 bc. Flax was grown and apparently woven into textiles early in the Neolithic Period.

The transition from hunting and food gathering to dependence on food production was gradual, and in a few isolated parts of the world this transition has not yet been accomplished. Crops and domestic meat supplies were augmented by fish and wildfowl as well as by the meat of wild animals. The farmer began, most probably, by noting which of the wild plants were edible or otherwise useful and learned to save the seed and to replant it in cleared land. Lengthy cultivation of the most prolific and hardiest plants yielded stable strains. Herds of goats and sheep were assembled from captured young wild animals, and those with the most useful traits—such as small horns and high milk production—were bred. The wild aurochs was the ancestor of European cattle, and an Asian wild ox of the zebu, was the ancestor of the humped cattle of Asia. Cats, dogs, and chickens were also domesticated very early.

Neolithic farmers lived in simple dwellings caves and small houses of sunbaked mud brick or reed and wood. These homes were grouped into small villages or existed as single farmsteads surrounded by fields, sheltering animals and humans in adjacent or joined buildings. In the Neolithic Period, the growth of cities such as Jericho (founded about 9000 bc) was stimulated by the production of surplus crops.

Pastoralism (individual country living) may have been a later development. Evidence indicates that mixed farming, combining cultivation of crops and stock raising, was the most common Neolithic pattern. Nomadic herders, however, roamed the steppes of Europe and Asia, where the horse and camel were domesticated.

The earliest tools of the farmer were made of wood and stone. They included the stone adz, an axlike tool with blades at right angles to the handle, used for woodworking; the sickle or reaping knife with sharpened stone blades, used to gather grain; the digging stick, used to plant seeds and, with later adaptations, as a spade or hoe; and a rudimentary plow, a modified tree branch used to scratch the surface of the soil and prepare it for planting. The plow was later adapted for pulling by oxen.

The hilly areas of southwestern Asia and the forests of Europe had enough rain to sustain agriculture, but Egypt depended on the annual floods of the Nile River to replenish soil moisture and fertility. The inhabitants of the Fertile Crescent around the Tigris and Euphrates rivers in the Middle East also depended on annual floods to supply irrigation water. Drainage was necessary to prevent the erosion of land from the hillsides through which the rivers flowed. The farmers who lived in the area near the Huang He developed a system of irrigation and drainage to control the damage caused to their fields in the flood plain of the meandering river.

Although Neolithic settlements were more permanent than the camps of hunting peoples, villages had to be moved periodically in some areas when the fields lost their fertility from continuous cropping. This was most necessary in northern Europe, where fields were produced by the slash-and-burn method of clearing. Settlements along the Nile River, however, were more permanent, because the river deposited fertile silt annually.

Historical Agriculture Through the Roman Period

With the close of the Neolithic period and the introduction of metals, the age of innovation in agriculture was largely over. The historical period—known through written and pictured materials, including the Bible; Middle Eastern records and monuments; and Chinese, Greek, and Roman writings—was highlighted by agricultural improvements. A few high points must serve to outline the development of worldwide agriculture in this era, roughly defined as 2500 bc to ad 500. For a similar period of development in Central and South America, somewhat later in date

Some plants became newly prominent. Grapes and wine were mentioned in Egyptian records about 2900 bc, and trade in olive oil and wine was widespread in the Mediterranean area by the 1st millennium bc. Rye and oats were cultivated in northern Europe about 1000 bc.

Many vegetables and fruits, including onions, melons, and cucumbers, were grown by the 3rd millennium bc in Ur (now Iraq). Dates and figs were an important source of sugar in the Middle East, and apples, pomegranates, peaches, and mulberries were grown in the Mediterranean area. Cotton was grown and spun in India about 2000 bc, and linen and silk were used extensively in 2nd-millennium bc China. Felt was made from the wool of sheep in Central Asia and the Russian steppes.

The horse, introduced to Egypt about 1600 bc, was already domesticated in Mesopotamia and Asia Minor. The ox-drawn four-wheeled cart for farm work and two-wheeled chariots drawn by horses were familiar in northern India in the 2nd millennium bc.

Improvements in tools and implements were particularly important. Tools of bronze and iron were longer lasting and more efficient, and cultivation was greatly improved by such aids as the ox-drawn plow fitted with an iron-tipped point, noted in the 10th century bc in Palestine. In Mesopotamia in the 3rd millennium bc a funnel-like device was attached to the plow to aid in seeding, and other early forms of seed drills were used in China. Farmers in China further improved efficiency with the invention of a cast-iron moldbar plow. Threshing was also done with animal power in Palestine and Mesopotamia, although reaping, binding, and winnowing were still done by hand. Egypt retained hand seeding through this period on individual farm plots and large estates alike.

Storage methods for oil and grain were improved. Granaries—jars, dry cisterns, silos, and bins containing stored grain—provided food for city populations. Without adequate food supplies and trade in both food and nonfood items, the high civilizations of Mesopotamia, northern India, Egypt, Greece, and Rome would not have been possible.

Irrigation systems in China, Egypt, and the Middle East were refined and expanded, putting more land into cultivation. The forced labor of peasants and the growth of bureaucracies to plan and supervise work on irrigation systems were probably basic in the development of the city-states of Sumer (now Iraq and Kuwait). Windmills and water mills, developed toward the end of the Roman period, increased control over the many uncertainties of weather. The introduction of fertilizer, mostly animal manures, and the rotation of fallow and crop land increased crop production.

Mixed farming and stock raising, which were flourishing in the British Isles and on the continent of Europe as far north as Scandinavia at the beginning of the historical period, already displayed a pattern that persisted throughout the next 3,000 years. In many regions, fishing and hunting supplemented the food grown by farmers.

About ad 100 Roman historian Cornelius Tacitus described the Germans as a tribal society of free peasant warriors who cultivated their own lands or left them to fight. About 500 years later, a characteristic European village had a cluster of houses in the middle, surrounded by rudely cultivated fields comprising individually owned farmlands; and meadows, woods, and wasteland were used by the entire community. Oxen and plow were passed from one field to another, and harvesting was a cooperative effort.

The Roman Empire appears to have started as a rural agricultural society of independent farmers. In the 1st millennium bc, after the city of Rome was established, however, agriculture started a development that reached a peak in the Christian era. Large estates that supplied grain to the cities of the empire were owned by absentee landowners and cultivated by slave labor under the supervision of hired overseers. As slaves, usually war captives, decreased in number, tenants replaced them. The late Roman villa of the Christian era approached the medieval manor in organization; slaves and dependent tenants were forced to work on a fixed schedule, and tenants paid a predetermined share to the estate owner. By the 4th century ad, serfdom was well established, and the former tenant was attached to the land.

Feudal Agriculture

The feudal period in Europe began soon after the fall of the Roman Empire, reaching its height about ad 1100. This period was also marked by development of the Byzantine Empire and the power of the Saracens in the Middle East and southern Europe. Agriculture in Spain, Italy, and southern France, in particular, was affected by events outside continental Europe.

As the Arab influence extended to Egypt and later Spain, irrigation was extended to previously sterile or unproductive land. In Egypt, grain production was sufficient to allow the country to sell wheat in international markets. In Spain, vineyards were planted on sloping land, and irrigation water was brought from the mountains to the plains. In some areas of the Middle East, oranges, lemons, peaches, and apricots were cultivated.

Rice, sugarcane, cotton, and vegetables such as spinach and artichokes, as well as the characteristic Spanish flavoring saffron, were produced. The silkworm was raised and its food, the mulberry tree, was grown.

By the 12th century agriculture in the Middle East had become static, and Mesopotamia declined to subsistence production levels when irrigation systems were destroyed by invading Mongols. The Crusades, however, increased European contact with Islamic lands and familiarized western Europe with citrus fruits and silk and cotton textiles.

The structure of agriculture was not uniform. In Scandinavia and eastern Germany, the small farms and villages of previous years remained. In mountainous areas and in the marshlands of Slavic Europe, the manorial system could not flourish.

A manor required roughly 350 to 800 hectares (about 900 to 2,000 acres) of arable land and the same amount of other prescribed lands, such as wetlands, wood lots, and pasture.

Typically, the manor was a self-contained community. On it was the large home of the holder of the fief—a military or church vassal of rank, sometimes given the title lord—or of his steward. A parish church was frequently included, and the manor might make up the entire parish. One or more villages might be located on the manor, and village peasants were the actual farmers. Under the direction of an overseer, they produced the crops, raised the meat and draft animals, and paid taxes in services, either forced labor on the lord's lands and other properties or in forced military service.

A large manor had a mill for grinding grain, an oven for baking bread, fishponds, orchards, perhaps a winepress or oil press, and herb and vegetable gardens. Bees were kept to produce honey.

Woolen garments were produced from sheep raised on the manor. The wool was spun into yarn, woven into cloth, and then sewn into clothing. Linen textiles could also be produced from flax, which was grown for its oil and fiber.

The food served in a feudal castle or manor house varied according to the season and the lord's hunting prowess. Hunting for meat was, indeed, the major nonmilitary work of the lord and his military retainers. The castle residents could also eat domestic ducks, pheasants, pigeons, geese, hens, and partridges; fish, pork, beef, and mutton; and cabbages, turnips, carrots, onions, beans, and peas. Bread, cheese and butter, ale and wine, and apples and pears also appeared. In southern Europe olives and olive oil might be used, often instead of butter.

Leather was produced from the manor's cattle. Horses and oxen were the beasts of burden; as heavier horses were bred and a new kind of harness was developed, they became more important. A blacksmith, wheelwright, and carpenter made and maintained crude agricultural tools.

The cultivation regime was rigidly prescribed. The arable land was divided into three fields: one sown in the autumn in wheat or rye; a second sown in the spring in barley, rye, oats, beans, or peas; and the third left fallow. The fields were laid out in strips distributed over the three fields, and without hedges or fences to separate one strip from another. Each male peasant head of household was allotted about 30 strips. Helped by his family and a yoke of oxen, he worked under the direction of the lord's officials. When he worked on his own fields, if he had any, he followed village custom that was probably as rigid as the rule of an overseer.

About the 8th century a four-year cycle of rotation of fallow appeared. The annual plowing routine on 400 hectares would be 100 hectares plowed in the autumn and 100 in the spring, and 200 hectares of fallow plowed in June. These three periods of plowing, over the year, could produce two crops on 200 hectares, depending on the weather. Typically, ten or more oxen were hitched to the tongue of the plow, often little more than a forked tree trunk. The oxen were no larger than modern heifers. At harvest time, all the peasants, including women and children, were expected to work in the fields. After the harvest, the community's animals were let loose on the fields to forage.

Some manors used a strip system. Each strip, with an area of roughly 0.4 hectare (about 1 acre), measured about 200 m (about 220 yd) in length and from 1.2 to 5 m (4 to 16.5 ft) in width. The lord's strips were similar to those of the peasants distributed throughout good and bad field areas. The parish priest might have lands separate from the community fields or strips that he worked himself or that were worked by the peasants.

In all systems, the lord's fields and needs came first, but about three days a week might be left for work on the family strips and garden plots. Wood and peat for fuel were gathered from the commonly held wood lots, and animals were pastured on village meadows. When surpluses of grain, hides, and wool were produced, they were sent to market.

In about 1300 a tendency developed to enclose the common lands and to raise sheep for their wool alone. The rise of the textile industry made sheep raising more profitable in England, Flanders (now in Belgium), Champagne (France), Tuscany and Lombardy (Italy), and the Augsburg region of Germany. At the same time, regions about the medieval towns began to specialize in garden produce and dairy products. Independent manorialism was also affected by the wars of 14th- and 15th-century Europe and by the widespread plague outbreaks of the 14th century. Villages were wiped out, and much arable land was abandoned. The remaining peasants were discontented and attempted to improve their conditions.

With the decline in the labor force, only the best land was kept in cultivation. In southern Italy, for instance, irrigation helped increase production on the more fertile soils. The emphasis on grain was replaced by diversification, and items requiring more care were produced, such as wine, oil, cheese, butter, and vegetables.

Scientific Agriculture

By the 16th century, population was increasing in Europe, and agricultural production was again expanding.

The nature of agriculture there and in other regions was to change considerably in succeeding centuries. Several reasons can be identified for this trend. Europe was cut off from Asia and the Middle East by an extension of Ottoman power. New economic theories were put into practice, directly affecting agriculture. Continued wars between England and France, within each of these countries, and in Germany consumed capital and human resources.

A new period of global exploration and colonization was undertaken to circumvent the Ottoman Empire's control of the spice trade, to provide homes for religious refugees, and to provide new resources for European nations convinced that only precious metals constituted wealth.

Colonial agriculture was intended not only to feed the colonists but also to produce cash crops and to supply food for the home country. This meant cultivation of such crops as sugar, cotton, tobacco, and tea, and production of animal products such as wool and hides.

From the 15th to the 19th century the slave trade provided laborers needed to fill the large workforce required by colonial plantations. Many early slaves replaced indigenous peoples who died from diseases carried by the colonists or were killed by hard agricultural labor to which they were unaccustomed. Slaves from Africa worked, for example, on sugar plantations in the Caribbean region and on indigo and cotton plantations in what would become the southern United States. Native Americans were virtually enslaved in Mexico. Indentured slaves from Europe, especially from the prisons of Great Britain, provided both skilled and unskilled labor to many colonies. Both slavery and serfdom were substantially wiped out in the 19th century. See Peonage; Plantation; Slavery.

When encountered by the Spanish conquistadors, the more advanced Native Americans in the New World—the Aztec, Inca, and Maya—already had intensive agricultural economies, but no draft or riding animals and no wheeled vehicles. Squash, beans, peas, and corn had long since been domesticated. Land was owned by clans and other kinship groups or by ruling tribes that had formed sophisticated governments, but not by individuals or individual families. Several civilizations had risen and fallen in Central and South America by the 16th century.

The scientific revolution resulting from the Renaissance and the Age of Enlightenment in Europe encouraged experimentation in agriculture as well as in other fields. Trial-and-error efforts in plant breeding produced improved crops, and a few new strains of cattle and sheep were developed. Notable was the Guernsey cattle breed, which is still a heavy milk producer. Land enclosure was increasingly practiced in the 18th century, enabling individual landowners to determine the disposition of cultivated land and pasture that previously had been subject to common use.

Crop rotation, involving alternation of legumes with grain, was more readily practiced outside the village strip system inherited from the manorial period. In England, where scientific farming was most efficient, enclosure brought about a fundamental reorganization of land ownership. From 1660 large landowners had begun to add to their properties, frequently at the expense of small independent farmers. By the mid-19th century the agricultural pattern was based on the relationship between the landowner, dependent on rents; the farmer, producer of crops; and the landless laborer, the hired hand of American farming lore. Drainage brought more land into cultivation, and, with the Industrial Revolution, farm machinery was introduced.

It is not possible to fix a clear decade or series of events as the start of the agricultural revolution through technology. Among the important advances were the purposeful selective breeding of livestock, begun in the early 1700s, and the spreading of limestone on farm soils in the late 1700s. Mechanical improvements in the traditional wooden plow began in the mid-1600s with small iron points fastened onto the wood with strips of leather. In 1797, Charles Newbold, a blacksmith in Burlington, New Jersey, reconceived of the cast-iron moldboard plow (first used in China nearly 2,000 years earlier). John Deere, another American blacksmith, further improved the plow in the 1830s and manufactured it in steel. Other notable inventions included the seed drill of English farmer Jethro Tull, developed in the early 1700s and progressively improved for more than a century; the reaper of American Cyrus McCormick in 1831; and numerous new horse-drawn threshers, cultivators, grain and grass cutters, rakes, and corn shellers. By the late 1800s, steam power was frequently used to replace animal power in drawing plows and in operating threshing machinery.

The demand for food for urban workers and raw materials for industrial plants produced a realignment of world trade. Science and technology developed for industrial purposes were adapted for agriculture, eventually resulting in the agribusinesses of the mid-20th century.

In the 17th and 18th centuries the first systematic attempts were made to study and control pests. Before this time, handpicking and spraying were the usual methods of pest control. In the 19th century, poisons of various types were developed for use in sprays, and biological controls such as predatory insects were also used. Resistant plant varieties were cultivated; this was particularly successful with the European grapevine, in which the grape-bearing stems were grafted onto resistant American rootstocks to defeat the Phylloxera aphid.

Improvements in transportation affected agriculture. Roads, canals, and rail lines enabled farmers to obtain needed supplies from remote suppliers and market their produce over a wider area. Food could be protected during transport more economically than before as the result of rail, ship, and refrigeration developments in the late 19th and early 20th centuries. Efficient use of these developments led to increasing specialization and eventual changes in the location of agricultural suppliers. In the last quarter of the 19th century, for example, Australian and North American suppliers displaced European suppliers of grain in the European market. When grain production proved unprofitable for European farmers, or an area became more urbanized, specialization in dairying, cheesemaking, and other products was emphasized.

The impetus toward increased food production following World War II (1939-1945) was a result of a new population explosion. A so-called green revolution, involving selective breeding of traditional crops for high yields, new hybrids, and intensive cultivation methods adapted to the climates and cultural conditions of densely populated countries such as India, temporarily stemmed the pressure for more food. A worldwide shortage of petroleum in the mid-1970s, however, reduced the supplies of nitrogen fertilizer essential for the success of the new varieties. Simultaneously, erratic weather and natural disasters such as drought and floods reduced crop levels throughout the world. Famine became common in many parts of Africa south of the Sahara. Economic conditions, particularly uncontrolled inflation, threatened the food supplier and the consumer alike. These problems became the determinants of agricultural change and development. See Energy Supply, World; Environment; Food Supply, World.

Industrial Agriculture

Many of the innovations introduced to agriculture by the scientific and Industrial revolutions paved the way for a qualitative change in the nature of agricultural production, particularly in advanced capitalist countries. This qualitative change became known as industrial agriculture. It is characterized by heavy use of synthetic fertilizers and pesticides; extensive irrigation; large-scale animal husbandry involving animal confinement and the use of hormones and antibiotics; reliance on heavy machinery; the growth of agribusiness and the commensurate decline of family farming; and the transport of food over vast distances. Industrial agricultural has been credited with lowering the cost of food production and hence food prices, while creating profitable businesses and many jobs in the agricultural chemistry and biotechnology industries. It has also allowed farmers and agribusinesses to export a large percentage of their crops to other countries. Farm exports have enabled farmers to expand their markets and have contributed to aiding a country's trade balance.

At the same time, industrial-scale agriculture has had adverse environmental consequences, such as intensive use of water, energy, and chemicals. Many aquifers and other water reservoirs are being drained faster than they can be renewed. The energy required to produce nitrogen-based synthetic fertilizers, to operate heavy farm equipment, to manufacture pesticides, and to transport food over long distances involves burning large amounts of fossil fuels, which in turn contribute to air pollution and global warming. The use of synthetic fertilizers has affected the ability of soil to retain moisture, thus increasing the use of irrigation systems. Fertilizer runoff has also stimulated algae growth in water systems. Finally, herbicides and insecticides in many cases have contaminated ground and surface waters.

During the 20th century, a reaction developed to industrial agriculture known as sustainable agriculture. While industrial agriculture aims to produce as much food as possible at the lowest cost, the main goal of sustainable agriculture is to produce economically viable, nutritious food without damaging natural resources such as farmland and the local watershed. Examples of sustainable agricultural practices include rotating crops from field to field to prevent the depletion of nutrients from the soil, using fertilizers produced naturally on the farm rather than synthetic products, and planting crops that will grow without needing extensive irrigation. Sustainable agricultural practices have seen great success in parts of the developing world where resources such as arable land and water are in short supply and must be carefully utilized and conserved.

THE ORIGINS OF AGRICULTURE

Recent archaeological finds place the beginning of agriculture before 7000 B.C. and animal domestication (mostly dogs used as hunting aids) thousands of years before that. There is some evidence that the people of Shanidar, in Kurdistan, were domesticating sheep and planting wheat as long ago as 9800 B.C.

Intensive food gathering, in which the local inhabitants of a region set up permanent residences and made extensive use of already present plants, seems to have started in the Near East around 9000 - 7000 B.C.

Barring the use of time machines, there is no way to know for sure how planting really got started. But archaeologists have lots of theories. One theory suggests that some seeds were spilled in a memorable manner during a migration. When the tribe next passed the same place, they might have correlated the spill of seeds with the sudden abundance of the plant. They could then have realized that they could store seeds and plant them, and be assured of having a food supply. later they began selecting and planting the seeds from plants with the highest yield. In this way,. plants were domesticated, changed and controlled to benefit man rather than just exist in the wild.

At about the same time as the agricultural advances described above, people started to domesticate the wild ox and gather sheep into herds. Remains of a hunting dog, dated back to 8500 B.C., have been found in North America.

TOWNS AND CITIES DEVELOP FROM FARMING

The abundance of the harvest from domesticated plants allowed major increases in population. Having all of one's plants and animals in one place allowed the agriculturist to move from random caves and makeshift huts into permanent or semi permanent villages with homes made from stones, wood, or wattle. An early example is the Biblical city of Jericho. It started as such a village around 9000 B.C., and has been a settlement of one sort or another ever since. One of the earliest recorded towns is Catal Huyuk established on the Konya Plain in Turkey. It is a vast, fertile expanse ideal for primitive agriculture. The earliest buildings date from 6500 B.C. and are similar to those found in the oldest Jericho settlements. You entered the mud brick buildings from the top. Catal Huyuk is notable for the number of shrines used for a variety of purposes, including burial and possible propitiation of deities of the hunt and the harvest. This implies an early religious organization and a way of life that left enough time for some members of the society to concentrate on religious duties. There was also time for crafts. some of the earliest known pottery was found in Catal Huyuk. There is also evidence of copper smithing and rope making, and some ovens were big enough to imply that some residents were full time bakers.

By 5000 B.C., the Euphrates Valley was full of villages and townships. The townships provided central services of storage, religious observance and administration that the villages could not handle. These townships developed into the Sumerian civilization.

At about the same time, similar villages were beginning in the Nile Valley and the river valleys of china and India.

EARLY FARMING TECHNIQUES

The initial approach to farming was to remove some of the seeds from food plants before eating them, then scatter the seeds back into the same area they came from.

Later, the planters realized that other (non -food) plants were competing with their plants for the field, so they took to weeding the fields to make sure the only their plants were growing there. Everything else was left to nature.

Eventually it became obvious that this constant replanting resulted in stunted crops and low yields. The first response was simply to find a new field. After all, the land was vast and people were few. After awhile, though, the obvious fields were used up. Then potential farmers looked to the forests.

Slash and Burn

Most agricultural societies discovered the slash and burn technique. First, all the foliage in a section of a forest was cut down, creating a field. The remains were left on the ground. Then the field was set on fire, and the ash from the cut foliage enriched the soil. After many uses even this enriched soil became barren, and farmers were forced to find new fields. As the population of the world grew and more fields were slashed and burned, the walk to a newly burned field became longer and longer and other cultures could claim these unattended fields. The tribe would then have to move to new sections of forest. In some areas, such as Madagascar, slash and burn agriculture is still practiced and the land is becoming less and less fertile.

Fallow Fields

A fallow field is one that is not planted for a period in hopes that it will regain its fertility. It is believed that the practice of leaving fields fallow originated because some cultures were forced to return to their old fields, and found that the infertile fields they left behind had become more productive.

This led to the establishment of a rotation system where each growing season certain fields would be left alone or tilled but not planted, extending the useful production life of a set number of fields. sometimes the fallow fields were used for pasturage for animals, which had the incidental benefit of fertilizing the soil.

It was later found that certain plants, thought useless except perhaps for animal fodder, were beneficial to a field's productivity, and seeds for these plants were planted in fallow fields.

Irrigation

As populations grew and competed for the best growing lands some cultures were forced to try to farm normally arid areas. Some of these cultures died trying; others discovered the principles of irrigation. There were some early massive engineering projects to dam water for later use, including the digging of canals to distribute water to normally dry fields. The first known examples of this irrigation process were built by farmers who colonized the Euphrates River Valley around 4000 B.C.

In most cases, irrigation involves trapping and storing water that appears for a short period, such as the spring flooding of the Euphrates and Nile, or the winter rainstorms of the American desert, so that it can be used later in normally dry periods. In almost all cases, early irrigation made the desert flower for a couple of centuries, then the water dried up in some climatic change or the fields grew barren because the irrigation had washed away all the good soil and the culture died. Both the Pueblo dwellers of the American desert and the inhabitants of Petra in the Middle East flourished and then died with their irrigation systems.

Other areas, such as the very fertile Nile Valley and the Tigris-Euphrates Fertile Crescent, were big enough and had a sufficiently dependable source of water so that they remained productive until the present day, though even these areas have undergone a decline in fertility and might be barren if not for modern agricultural techniques.

STORY OF FARMING

"Over 80 percent of mankind's diet is provided by the seeds of less than a dozen plant species." Over the years man has invented new machines and techniques to increase the amount and variety of crop production. The following will be an overview of the history of farming. We will examine the major historical cultures, the development of the tractor, and the major types of agriculture practiced today. The roots of farming began in the areas of present day Turkey and the Middle East about 10,000 years ago. Two of the earliest settlements are known as Catal Hüyük and Jericho. Catal Hüyük had, by 6000 B.C., more then 1000 houses. It is at this place that we have discovered evidence of people taking wild grasses and using the seeds for food and planting for the next years food. These seeds are now known as cereals and make up a large percentage of the worlds food supply. Jericho, like many early cities, was located around a consistent water source, a spring which produced over 1000 gallons of water every minute. Jericho consisted of about eight to ten acres on which it is estimated that two to three thousand people lived. These people were supported by farming of wheat, barley, peas, and lentils. Archeologist believe the earliest settlers in this area were a small group of hunter-gathers. Hunter - gathers would live off the land forging berry and edible plants, as well as hunting wild animals.

These types of people lived in smaller groups because they had to be mobile to find more food. It was not until man began to plant and harvest crops that large permanent settlements could be established, like at Jericho. We find many of the early civilizations began along major river systems. For example Egyptians settled along the Nile River, Harappa culture along the Indus, Chinese Empire along the Huang River and the Mesopotamian Countries along the Tigris and Euphrates rivers. The river systems provided these early civilizations with a consistent source of silt from the yearly floods and water for the crops. The silt is like a natural fertilizer, bringing new minerals to enrich the crop depleted soil. Farming changed very little from early times until about 1700. In the 1700's an agriculture revolution took place which led to a large increase in the production of crops. This increase of crops came about in a large part by "... little more then the final destruction of medieval institutions and the more general adoption of techniques and crops which had been know for a long time".

Included in some of these changes was also the adoption of crops from the "new world" such as corn and potatoes which produced a very large yield.

In the 1850's, the industrial revolution spilled over to the farm with new mechanized methods which increased production rates. Early on, the large changes were in the use of new farm implements. Most of these early implements were still powered by horse or oxen. These new implements combined with crop rotation, manure and better soil preparation lead to a steady increase of crop yield in Europe. The advent of steam power and later gas powered engines brought a whole new dimension to the production of crops. Yet, even as recent as 100 years ago, four-fifth of the world populations lived outside towns and were in some way dependant on agriculture. Even in 1970's Griggs suggests that half of the worlds working population is still employed in agriculture. In the following pages we will examine the cultures, the farming types, the tools, and much more as they relate to agriculture.

THE PLOW

The earliest plows where forked sticks and timbers. In the middle east the early plows were called ard. The early plows simply loosened the soil. A type of ard is still used in some underdeveloped countries today.

The major advance before 1000 A.D. was the development of the heavy plough, which was more than the simple plows farmers used earlier. It had a coulter which cut a thin strip in the turf. The coulter was followed by a share which would slice into the soil and then the soil would ride up the mouldboard which would turn it over. Later wheels were attached to this type of plow and later still a seat was added. By turning over the soil weeds were limited and overall it helped the growing process. Metal was added to parts of the plow which increased it's efficiency. One of the major problems was that the dirt would become stuck on the plow and had to be cleaned off by hand. A second problem was that this system did not work in the dense grasses of the western plains.

The problem of the plains was solved by a black smith from Vermont named John Deere. Deere moved to Grand Detour, Illinois in 1836. He invented a blade which was self polishing and combined the share and moldboard into a one piece plow. Deere moved his factory to Moline Illinois and began manufacturing in 1847. The blade was an amazing hit and began the John Deere company. Early farming utilized oxen any kind of cattle used for draft, or pulling work, are called oxen" in the fields. These animals appear to be first used around 3500 B.C. with primitive plows made of wood. In Europe the invention of the horse collar and shoe in the 9th century allowed the plow to be pulled by horse. Yet even into the 18th century oxen still outnumber the horses partly due to the expense of feeding the horse. Yet with the advent of the iron plows many farmers changed over to heavy horses which could pull the new type of implements at a faster pace than the oxen.

The Auto Plow

Early on in tractor design was the auto-plow. The Auto-Plow was a tractor with the plow blades already mounted to the bottom of the tractor. Left is a 1912 Hackney Auto Plow. It is believed to be the only running one of it's kind housed at the Martha and Dale Hawk Museum in North Dakota. Today plows are pulled by large tractors and can do large tracks of land in a single day.

CULTIVATING THE SOIL

After the fields were plowed, the seed bed had to be prepared. The better the soil was broken up, the better the crops would grow in the soil. Early seed beds were prepare by hand with sticks, spades, or rakes. An implement called a harrow was then used to brake up big clumps of soil. The early harrows were square shaped with spikes attached to a wooden frame pulled by a horse or ox. Later the design was changed to triangular, which made it easer to pull though the soil by the horse or ox. Around the eighteenth century cultivators began to take over from the harrows to work the soil.

Cultivators were then mounted on wheels which gave them a great advantage to the early models. These machines could control the depths at which they went into the soil. Modern Day cultivators are very large implements which need large tractors to pull them.

Sowing Seeds

Early planting was done by hand. The seeds would be thrown, or broadcast. This system made it more difficult to weed and harvest the crop. Later a dibber was used for some crops. A dibber was a board with holes evenly spread apart. A stick would be pushed through the holes and then a seed would be placed in the hole made by the stick. This was very effective but also very tedious and time consuming. The idea for dropping seeds through a tube first appeared in Mesopotamia about 1500 B. C. In 1701 Jethro Tull invented the first seed drill. The implement would cut small channels into the soil and the seed would be dropped into the channel. Before this time seeds were usually planted by a method known as broadcasting. Broadcasting is simply throwing seeds onto the ground. The seed drill had many advantages to the broadcasting system. First a much higher percentage of seed came to produce crops. Less seed was lost to birds or other animals. Finally, with rows, it was much easier for the farmer to weed his crop. Jethro Tull's invention was met with skepticism and not really appreciated or accepted till after his death in 1741.

REAPING

Early reaping was all done by hand. Reaping is the cutting of the grain. In Egypt a flint blade was used to cut the wheat. In Europe the scythe had been introduced by the Romans. Yet the Europeans continued to use the sickle until limited labor forced them to use the more efficient scythe. By hand a worker could cut about 0.3 acres in a day. An experiment with an old sickle harvested 6.25 pounds in one hour and was two pounds after being threshed. After being cut the stalks were tied into bundles and then let to dry. After drying the wheat would be threshed and winnowed. The first evidence of a machine reaper come from the Gauls in Europe. The Cradlers allowed the the cutter to deposit the shafts in a pile after the swing. They could mow 1.5 to 3 acres in a day. Later, labor shortage, both in Europe and especially in the Western United States, spurred the farmer on to find new and more efficient ways to harvest his crop. The first successful reaper was created by Rev. Patrick Bell in the early 1800's. In this design the reaper was pushed by horses with the shears cutting the wheat in front. The Bell reaper could cut ten acres a day and needed sharpening after fifty acres.

One of the largest used early reapers was one made by the McCormack company. The McCormack reaper was widely used and accepted in United States and England. The rest of Europe was much slower to adopt the new technology. In 1890 only one tenth of France or Germany had adopted the use of the reaper in their fields. The reaper would cut the stalks where they would lie in the fields until they were manually pick up. Later a board was added behind the blades and a man would push the stalks off into piles. The sail reaper came about in 1862. It now replaced a person pushing off the stalks by sail like rakes. The rakes would rotate and clean the stalk off into piles off the side of the machine. Then men would again manually bind up the stalk into bundles. It could harvest five acres in a day. The early reapers required a man to push the wheat off by hand.The reaper was improved to both reap and bind the stalks in later versions by McCormick and other manufactures. McCormack started mass-production of the reaping machines which bound the sheaves in 1877. Other manufactures also came out with their own versions of self binding machines.

THRESHING AND WINNOWING

Threshing is separating the grain from the stalks. In early days this was accomplished by men hitting it with a flail. After the wheat was separated, it would be tossed into the air to separated it from the chaff known as winnowing. In some countries the grain was spread on the floor and threshed by animal pulled heavy sleds drawn over the grains. After the grain is separated from the straw it would be again winnowed. This process could take up to two months. In early colonial times George Washington experimented with a new method of threshing. He built on his farm a round barn (first in America) and would stack wheat 3 feet high. Then the workers would take special trained horses and have them run on the wheat in circles in the barn. The wheat would be dislodged from the stalk by the horses hooves and would drop through spaces in the floor to a storage chamber down below it. Another famous American, Thomas Jefferson, imported a threshing machine from Europe and was able to thresh 120 bushels of grain per day.

THRESHING MACHINES

In 1786, Scotsman Andrew Meikle invented a threshing machine. The early machines were powered by horse or later by steam engines. Many of them were stationary. The threshing machines did four different tasks. It removed grain, separated the grain from the cob or husk, cleaned the grain, and then gathered it or stacked it. The threshing machine was never widely purchased by small farmers. They were very large and often too expensive for the average farmer. Often they were used in custom operation going from farm to farm. Later these threshing machines were combined with reapers and became known as combines.

COMBINES

The first combine was invented by Hiran Moore in 1838. It took several decades before the combine came into wide use. Early combines were driven by as many as 16 or more horses. Later they were pulled by steam engine and then combined into a single machine by George Stockton Berry. Berry took the straw and used it for fuel to heat the boiler. The header (or cutter) was over forty feet. This machine could cut and thresh over one hundred acres in a day. The cost of the machine was also cheaper than the horse drawn reapers and stationary threshers. The cost of the reaper and thresher was about $3 a acre while the combine was between $1.50 and $1.75. The modern combine has many luxuries. Most are now equipped with a full stereo system, comfortable seat, and full air conditioning. The inside of the cab is slightly pressurized so that the air pushes out of the cab which does not allow dust and dirt to come in.

STEAM ENGINES

Steam engines were invented in the late 1700's and applied to moving vehicles by the early 1800's. Later, it's application was put to farming equipment. The application of steam engines was limited because of the enormous weight that was required in the machine. "The introduction of high-pressure boilers in the 1850's did much to lighten engines." The steam engine enjoyed it's largest amount of use between 1885 and 1914.

The steam engine was first applied in Europe to the threshing process and to drainage pumps. It was not until the 1850 that the Steam Engine first was used in plowing in Europe. "The steam plough, although able to plough ten times the area that horses could plough in a day, was cumbersome and costly, and had only a limited impact on farming in either Europe or the United States. Thus the horse remained the mains source of power until the early twentieth century." "The most successful early application of steam in farming was to plowing. Before steam engines were self-propelling, and had to be hauled into position by horses, schemes for using them to haul ploughs across fields by cables had been devised." Steam engines had their drawbacks. First, boiler explosions were frequently caused by low water and other factors. The Steam engine was also very heavy and often would collapse bridges originally designed for simple horse and carriage.

INTERNAL COMBUSTION TRACTORS

The earliest internal combustion engines were stationary. The would run barn equipment like threshers and other machinery. Some were mounted on wheels so they could be transfered from field to field. Early tractors utilized wide metal tires, especially in the rear of the machine to disperse the weight. Front wheels often had ridges to help them steer in the dirt. Problems with traction pushed engineers to come to another form of wheels. A continuous belt with slats were fitted to the front and back wheels. This style has been made famous by Caterpillar company and is still used in heavy earth moving equipment today. Fordson was one of the first mass produced tractors starting in 1916. Plowing speed was 2.8 mph weighing over a ton. It ran on kerosene and could plow 8 acres on one tank of fuel. In 1932 Allis-Chalmers began to use pneumatic tires from Firestone Tire and Rubber Company. The tires had a much better grip in the soils. They have many advantages over the metal tires, including their weight. Today most of the tires used are wide and grooved for best results. Today tractors come in all sizes and are used for a multitude of tasks on the farm.

Fig. Internal Combustion Tractor

WORLD AGRICULTURE

Over the 10,000 years since agriculture began to be developed, peoples everywhere have discovered the food value of wild plants and animals, and domesticated and bred them. The most important crops are cereals such as wheat, rice, barley, corn, and rye; sugarcane and sugar beets; meat animals such as sheep, cattle, goats, and pigs or swine; poultry such as chickens, ducks, and turkeys; animal products such as milk, cheese, and eggs; and nuts and oils. Fruits, vegetables, and olives are also major foods for people. Feed grains for animals include soybeans, field corn, and sorghum.

Agricultural income is also derived from nonfood crops such as rubber, fiber plants, tobacco, and oil seeds used in synthetic chemical compounds, as well as animals raised for pelts. Conditions that determine what is raised in an area include climate, water supply and waterworks, terrain, and ecology. In 2003, 44 percent of the world's labor force was employed in agriculture. The distribution ranged from 66 percent of the economically active population in sub-Saharan Africa to less than 3 percent in the United States and Canada. In Asia and the Pacific the figure was 60 percent; in Latin America and the Caribbean, 19 percent; and in Europe, 9 percent. Farm size varies widely from region to region. In the early 2000s the average for Canadian farms was about 273 hectares per farm; for farms in the United States, 180 hectares. By contrast, the average size of a single land holding in India was 2 hectares.

Size also depends on the purpose of the farm. Commercial farming, or production for cash, usually takes place on large holdings. The latifundia of Latin America are large, privately owned estates worked by tenant labor. Single-crop plantations produce tea, rubber, and cocoa. Wheat farms are most efficient when they comprise thousands of hectares and can be worked by teams of people and machines. Australian sheep stations and other livestock farms must be large to provide grazing for thousands of animals. Individual subsistence farms or small-family mixed-farm operations are decreasing in number in developed countries but are still numerous in the developing countries of Africa and Asia. Nomadic herders range over large areas in sub-Saharan Africa, Afghanistan, and Lapland; and herding is a major part of agriculture in such areas as Mongolia. Much of the foreign exchange earned by a country may be derived from a single agricultural commodity; for example, Sri Lanka depends on tea, Denmark specializes in dairy products, Australia in wool, and New Zealand and Argentina in meat products. In the United States, wheat, corn, and soybeans have become major foreign exchange commodities in recent decades. The importance of an individual country as an exporter of agricultural products depends on many variables. Among them is the possibility that the country is too little developed industrially to produce manufactured goods in sufficient quantity or technical sophistication. Such agricultural exporters include Ghana, with cocoa, and Myanmar (formerly Burma), with rice. However, a developed country may produce surpluses that are not needed by its own population; this is the case with the United States, Canada, and some other countries. Because nations depend on agriculture not only for food but for national income and raw materials for industry as well, trade in agriculture is a constant international concern. It is regulated by the World Trade Organization. The Food and Agriculture Organization of the United Nations (FAO) directs much attention to agricultural trade and policies. According to the FAO, world agricultural production, stimulated by improving technology, grew steadily from the 1960s to the 1990s. Per capita food production saw sustained growth in Latin America, the Caribbean, Asia, and the Pacific, and limited growth in the Near East and North Africa. The only region not to experience growth during the 1980s and 1990s was sub-Saharan Africa, which suffered from climatic conditions that made agriculture difficult. Although agricultural growth began to taper off in the year 2000, it continued to outpace world population growth.

AGRICULTURE IN THE UNITED STATES

In North America, agriculture had progressed significantly before European colonists arrived. There is evidence that corn, or maize, was cultivated at least as early as 3,000 years ago in the southwestern United States. Although few Native Americans relied on domesticated animals, some groups had advanced methods of cultivating food crops. The Wampanoag peoples of what is now Massachusetts, for example, fertilized their corn seeds by burying fish in the ground near the seeds. The Iroquois of the eastern United States exploited the natural relationship between plants to make their crops more productive. They planted corn, beans, and squash together in small groups, so that the corn plants supported the beans, the nitrogen released by the roots of the bean plants fertilized the corn, and the sprawling squash vines reduced the number of weeds. Corn, beans, squash, potatoes, tomatoes, peanuts, cacao (chocolate), and many other plants were originally domesticated by Native Americans. Until the 19th century, agriculture in the United States shared the history of European and colonial areas and was dependent on European sources for seed, stocks, livestock, and machinery, such as it was. That dependency, especially the difficulty in procuring suitable implements, made American farmers somewhat more innovative. They were aided by the establishment of societies that lobbied for governmental agencies of agriculture; the voluntary cooperation of farmers through associations; and the increasing use of various types of power machinery on the farm. Government policies traditionally encouraged the growth of land settlement. The Homestead Act of 1862 and the resettlement plans of the 1930s were the key agricultural legislative acts of the 19th and 20th centuries. In the 20th century steam, gasoline, diesel, and electric power came into wide use. Chemical fertilizers were manufactured in greatly increased quantities, and soil analysis was widely employed to determine the elements needed by a particular soil to maintain or restore its fertility. The loss of soil by erosion was extensively combated by the use of cover crops (quick-growing plants with dense root systems to bind soil); contour plowing in which the furrows follow the contour of the land and are level, rather than running up and down hills and providing channels for runoff water; and strip cropping (sowing strips of dense-rooted plants to serve as water-breaks or windbreaks in fields of plants with loose root systems.

Selective breeding produced improved strains of both farm animals and crop plants. Hybrids of desirable characteristics were developed; especially important for food production was the hybridization of corn in the 1930s. New uses for farm products, by-products, and agricultural wastes were discovered. Standards of quality, size, and packing were established for various fruits and vegetables to aid in wholesale marketing. Among the first to be standardized were apples, citrus fruits, celery, berries, and tomatoes.

Improvements in storage, processing, and transportation also increased the widespread marketability of farm products. The use of cold storage warehouses and refrigerated railroad cars was supplemented by the introduction of refrigerated motor trucks, rapid delivery by airplane, and the quick-freeze process of preservation in which farm produce is frozen and packaged the same day that it is picked. Freeze-drying and irradiation have also reached practical application for many perishable foods. Scientific methods are now applied to pest control, limiting overuse of insecticides and fungicides and employing more varied and targeted application techniques. New understanding of significant biological control measures and the emphasis on integrated pest management make possible more effective control of certain kinds of insects. Chemicals for weed control are important for a number of crops, such as cotton and corn. The increasing use of chemicals for the control of insects, diseases, and weeds, however, has resulted in additional environmental problems and regulations that place strong demands on the skill of farmers.

Since the 1970s high technology farming, including new hybrids for wheat, rice, and other grains, better methods of soil conservation and irrigation, and the growing use of improved fertilizers has led to the production of more food per capita, not only in the United States, but in much of the rest of the world. United States farmers, however, still have the advantage of superior private and government research facilities to produce and perfect new technologies.

New applications of technologies at the beginning of the 21st century are further improving crop production. Precision farming, also known as prescription farming, site-specific farming, or variable rate farming, utilizes global positioning systems (GPS) and geographic information systems (GIS) in the satellite collection and transmission of data to create yield maps of fields during harvest. Farmers use the yield maps as they plant and fertilize their crops the following season. This increases crop production while reducing the use of both fertilizers and fuel. GPS also helps farmers comply with environmental regulations when applying fertilizers and pesticides.

Biotechnology is also increasing agricultural productivity. In recent years farmers have begun producing a new, genetically engineered oil seed crop that grows from canola, an oil seed producing plant, to yield lauric oil, which comes naturally from coconuts and palm kernels. New hybrid corn seed recently developed to resist the corn borer, an insect poisonous to that crop, and improved varieties of barley with disease resistant genes are now under cultivation, as well. Biotechnology developments have also become increasingly controversial, however, because it is difficult to determine the environmental consequences of genetically engineered organisms. Some people, including some scientists, object to any procedure that changes the genetic composition of an organism. Critics are concerned that some of the genetically altered forms will eliminate existing species, thereby upsetting the natural balance of organisms.

Farming Regions

The United States has ten major farming areas. They vary by soil, terrain, climate, and distance to markets and storage and marketing facilities. The states of the Northeast and the Great Lake states are the country's principal milk-producing areas. Climate and soil there are suited to raising grains and forage for cattle and for pastures. Chicken production is important to Maine, Delaware, and Maryland. Fruits and vegetables are also important to the region.

The Appalachian region is the major tobacco-producing area of the nation. Peanuts, cattle, and dairy production also are important. Beef cattle and chickens are the major livestock products in the southeastern states; fruits and vegetables and peanuts are also grown. Florida has vast citrus groves and winter vegetable production areas.

In the Mississippi Delta states, principal cash crops are soybeans and cotton. Rice and sugarcane are grown in the more humid and wet areas. With improved pastures, livestock production has gained importance in recent years. It also is a major chicken-producing region. The Corn Belt, extending from Ohio through Iowa, has rich soil, good climate, and sufficient rainfall for excellent farming. Corn, beef cattle, hogs, and dairy products are of primary importance. Other feed grains, soybeans, and wheat also are grown.

The northern and southern states of the Great Plains, extending from Canada to Mexico and from the Corn Belt to the Rocky Mountains, are restricted by low rainfall in the western portion and by cold winters and short growing seasons in the north. But about 60 percent of the nation's winter and spring wheat is grown in this region. Other small grains, grain sorghums, hay, forage crops, and pastures help make cattle important in these states. Cotton is produced in the southern area. The Rocky Mountain states provide yet a different terrain. Vast areas are suited to cattle and sheep. Wheat is important in the north. Irrigation in the valleys provides water for hay, sugar beets, potatoes, fruits, and vegetables. The Pacific region includes California, Oregon, and Washington, plus Alaska and Hawaii. In the northern mainland, farmers raise wheat, apples and other fruit, and potatoes. Dairy farming, vegetables, and barley are important to Alaska. Many farmers in the southern part of the region have large tracts on which they raise vegetables, fruit, and cotton, usually with irrigation. Cattle are raised throughout the region. Hawaii grows sugarcane and pineapple as its major crops.

Agricultural Resources

The total land area of the United States is 916.2 million hectares (2,264 million acres), of which 42 percent is used to produce crops and livestock. The rest is distributed among forest land (33 percent) and urban, transportation, and other uses. At the beginning of the 21st century, cropland resources comprised 174 million hectares (431 million acres). About 82 percent of cropland is cultivated, including 32 million hectares (79 million acres) used for corn, 26 million hectares (64 million acres) used for hay, and 25 million hectares (61 million acres) used for wheat. More than 50 percent of croplands are prime farmland, the best land for producing food and fiber.

The nation has another nearly 400 million hectares (almost 1 billion acres) of nonfederal rural land currently being used for pastures, range, forest, and other purposes. About 27.5 million hectares (about 68 million acres) of this land are suitable for conversion to cropland if needed.

Recent Changes

The history of agriculture in the United States since the Great Depression has been one of consolidation and increasing efficiency. From a high of 6.8 million farms in 1935, the total number declined to 2.1 million by 2005. In 2005 the area devoted to farms occupied 378 million hectares (933 million acres). Average farm size in 1935 was about 63 hectares (about 155 acres); in 2005 it was 180 hectares (444 acres). About 3.9 million people lived on farms according to the 2000 U.S. census, based on a farm definition introduced in 1977 to distinguish between rural residents and people who earned $1,000 or more from annual agricultural product sales. The farm population continues to constitute a declining share of the nation's total; about 1 person in every 72, or 1.4 percent of the nation's 281 million people in 2000, were farm residents. Total value of land and buildings on U.S. farms in 2002 was $1.19 trillion. The value of products sold was about $200 billion. Overall net farm income reached a record high of $85.5 billion in 2004, of which government subsidies accounted for 14.6 percent. Not including real estate, major expenditures by farmers in 2002 were for feed ($31.7 billion); purchases of livestock and poultry ($27.4 billion); fertilizer, chemicals, and seeds ($25 billion); hired labor ($22 billion); and fuel ($6.7 billion). Outstanding farm debt in 2002 was $193.3 billion, of which 53 percent was owed on real estate. Interest payments on the mortgage debt were about $8.6 billion per year. In 1980 a report based on projections by the U.S. government stated that in the next 20 years world food requirements would increase tremendously, with developed countries requiring most of the increase, and that food prices would double. Less than five years later, however, the U.S. farmer was enveloped in a major crisis caused by exceptionally heavy farm debts, mounting farm subsidy costs, and rising surpluses. A number of farmers were forced into foreclosure. The ailing Farm Credit System, a group of 37 farmer-owned banks under the Farm Credit Administration appealed to the government for a $5 to $6 billion fund that would keep the system solvent despite the weak national farm economy. After initial resistance, President Reagan signed legislation in December 1985 designed to create the Farm Credit System Capital Corporation to take over bad loans from the system's banks and to assume responsibility for foreclosing or restructuring distressed loans.

President Reagan also signed the Food Security Act of 1985, legislation designed to govern the nation's farm policies for the next five years, trim farm subsidies, and stimulate farm exports. In the early 1990s, although farms generally still struggled to be economically viable, fewer farms faced the kinds of crises prevalent in the 1980s. Government assistance to farms steadily decreased; for example, the number of acres placed in federal commodity programmes, which pay farmers to leave acres uncultivated, decreased by nearly 84 percent from 1987 to 1992.

Farm subsidies began to increase again between 1995 and 2003 in response to low commodity prices and natural disasters, with such legislation as the Freedom to Farm Act (1996) and the Farm Security and Rural Investment Act (2002). In all, Congress passed emergency farm assistance legislation five times between 1998 and 2001. Following these measures, the national farming debt-to-equity ratio decreased each year from 2002 to 2005.

AGRICULTURAL EXPORTS

The United States is the world's principal exporter of agricultural products, followed by Netherlands, France, Germany, Brazil, Belgium, and Italy. In 2004 the value of U.S. agricultural products exported was $64 billion.

A substantial percentage of the wheat, soybeans, rice, cotton, tobacco, and corn for grain produced in the United States is exported. The major foreign markets are Asia, Western Europe, and Latin America. Japan heads the list of individual countries that import U.S. farm products.

Chapter 2

Formation of Soil

Soil may be defined as a thin layer of earth's crust which serves as a natural medium for growth of plants. It is the unconsolidated mineral matter that has been subjected to, and influenced by, genetic and environmental factors— parent material, climate, organisms and topography all acting over a period of time. Soil differs from the parent material in the morphological, physical, chemical and biological properties. Also, soils differ among themselves in some or all the properties, depending on the differences in the genetic and environmental factors. Thus some soils are red, some are black; some are deep and some are shallow; some are coarse textured and some are fine-textured. They serve as a reservoir of nutrients and water for crops, provide mechanical anchorage and favourable tilth. The components of soil are mineral matter, organic matter, water and air, the proportions of which vary and which together form a system for plant growth.

The processes by which soil formation, or *pedogenesis*, occurs, are known collectively as *pedogenic processes*, of which there are four main groups—additions, transformations, transfers and losses. Additions involve both organic and mineral material, and can occur at the surface or within the soil itself. These materials are then transformed by the processes of organic matter decomposition, mineral weathering and clay mineral formation. Soil components can also undergo transfer from one part of the soil to another by a variety of processes which involve either transport in water or mechanical mixing. Some of the material will be lost from the soil via a number of processes, either as individual components or in combination with others. The pedogenic processes produce soil horizons, which combine to form soil profiles, different horizon combinations giving rise to different soil types. Each group of pedogenic processes in turn, although it is important to recognise that these processes rarely occur in isolation; soil profiles are products of the combined operation of a number of processes. The formation of soil horizons and a series of pedogenic pathways which produce basic types of soil profile. Finally, the classification of soil profiles will be discussed, looking at different methods of classification and their relative merits.

PEDOGENIC PROCESSES

Additions

In the context of soil formation, the material added to a soil can take two main forms—organic matter and mineral material—and addition can occur either at the surface or within the soil itself. Organic matter derived from above the ground includes material of local origin, supplied by vegetation growing in the soil or animals living at the surface, and also material derived from some distance away, being transported to the site by a variety of mechanisms. Locally derived material usually takes the form of plant litter and animal droppings, or larger but less frequent inputs when a whole organism dies, while transported material can include, for example, wind-blown leaves, stems carried by running water, a dead tree falling down a steep slope, or manure added to a soil to assist cultivation. Mineral material can also be added in a similar way, being blown or washed onto a soil, moved downslope under gravity or added by human activity which could include, for example, agrochemical application, refuse disposal or the movement of soil material during construction. Mineral material can also be carried by water in a dissolved form and therefore added via precipitation or water moving downslope. Other sources of mineral material include volcanic ash and glacial or marine sediments, although the volume of such materials may cause complete burial of the soil rather than additions to the profile.

Like above-ground inputs, subsurface additions can be locally derived or transported from elsewhere. Organic inputs of local derivation comprise plant roots, soil fauna and micro-organisms, while more distant material can be carried in solution or as small particles in water draining laterally through the soil. Mineral material can also be carried from some distance away by lateral drainage.

Transformations

Once organic material has been added to a soil, it will undergo decomposition, unless this is prevented by environmental circumstances or the material is already fully decomposed. Most organic matter decomposition relates to enzymic oxidation by organisms living in the soil or at its surface. As carbon and hydrogen make up the bulk of dry organic matter.

The breakdown of complex organic structures also leads to the formation of a variety of more simple, inorganic products. This process is known as *mineralisation,* and is an important source of plant nutrients such as nitrogen, sulphur and phosphorus; cations such as Ca^{2+}, Mg^{2+} and K^{+} are also released.

During organic matter decomposition, an additional set of processes occurs known as *humification*. These involve the complex reaction of various decomposition products to produce large, complex molecular chains, or *polymers*. These reactions result in the formation of a stable end-product known as humus, comprising small, colloidal particles, <2 ìm in size which, like clays, have a net negative charge, and are important in soil aggregation and cation exchange reactions. During one year of organic matter decomposition, approximately 60-80 per cent of the organic residues are oxidised and returned to the atmosphere as carbon dioxide, the rest remaining in the soil either as humus or in the bodies of soil organisms.

Two groups of compounds can be recognised in humus chemistry—humic and non-humic groups. Humic substances account for around 60-80 per cent of the soil organic matter and are characterised by aromatic, ring-type structures, which include polyphenols and polyquinones. These substances form by a variety of biological and chemical processes, the details and relative importance of which remain unclear. The non-humic group represents around 20-30 per cent of the soil organic matter and is generally less complex and less resistant to breakdown by microorganisms. It includes polysaccharides and polyuronides, along with organic acids and some protein-like substances. The components of humic material can be separated by chemical fractionation into three main groups—fulvic acid, humic acid and humin. These differ in their molecular weight, and the extent and strength of polymer bonding increases with molecular weight.

Mineral transformation can occur either by weathering or by the formation of clay minerals. Weathering is often considered to be of either a mechanical (physical) or chemical nature. Processes causing brittle fracture involve the application of mechanical stress or the release of strain, which can occur either as a single event or on a cyclical basis. The principal mechanisms which have been recognised are thermal expansion and contraction, freeze/thaw, salt crystal growth, root wedging and strain release. Although not always operative within soils themselves, these processes can be particularly important in the initial stages of parent material breakdown and in preparing this material for crystal lattice breakdown.

The effectiveness of thermal expansion and contraction, which occurs by diurnal heating and cooling, has been questioned, although there is general agreement that wetting and drying may increase its effectiveness in situations where the two processes are occurring simultaneously. Freeze-thaw, which operates as a result of water increasing its volume by about 9 per cent on freezing, produces three basic types of process—frost scaling, in which thin layers of ice form parallel to the rock surface; frost splitting, which occurs in massive rock; and frost wedging, which is probably dominant in most cases, in which cracks within a rock are exploited. Again, its effectiveness as a weathering mechanism has been questioned, although this remains open to debate because the processes involved are incompletely understood. Salt crystal growth can occur for three main reasons—evapouration of the salt solution, decreased solubility due to falling temperature, or mixing of two different salt solutions with the same major cation. Although some salts cause little breakdown, others, such as calcium chloride and magnesium sulphate can result in significant breakdown. Root wedging is the process whereby plant roots extend into rock crevices and exert pressures on the rock as they extend and widen, although its effectiveness is probably confined to weakly cohesive materials. In contrast, strain release is a process capable of producing major rock fracture. This occurs when an overburden is removed from a rock by erosion, causing it to rebound elastically, fracturing perpendicular to the direction of unloading. As this often results in sheets of fractured rock lying roughly parallel to the ground surface, the process is also known as *onion-skin* weathering or *exfoliation*.

Although a wide range of complex processes are involved in crystal lattice breakdown, these are often assigned, albeit simplistically, to a number of basic categories—solution, carbonation, hydration, hydrolysis, oxidation/ reduction and organic complexing.

Transfers

Water can transport material within a soil either in solution or suspension. This process is known as *translocation,* although the term *leaching* is also commonly used to refer to the movement of material in solution. While transport usually occurs in a downwards direction, it can in some cases operate laterally or upwards. Material carried in soil water can be derived from material which forms direct additions to the soil, or which is the product of transformation. Whether it is carried in solution or suspension will depend on its size and solubility, and these factors also determine whether it is redeposited in the soil or lost from it.

Transfer in solution can occur in the form of soluble material produced during transformation processes. Solution can also occur by the presence of H^+ and Al^{3+} ions in the soil drainage water, which displace base cations from the soil exchange complex. The passage of mobile anions through the soil can also attract base cations into solution. For example, $HCO3^-$ anions result from the dissociation of carbonic acid.

Unless all the base cations are redeposited elsewhere in the soil, they will gradually be lost via the ground drainage waters and the soil will therefore become progressively more acidic. Transfer in solution can also occur as a result of the uptake of nutrients by plant roots; these can be stored in the plant or returned to the soil via the plant litter, therefore taking the form of a *nutrient cycle.*

Redeposition of material carried in solution will occur when a change in soil conditions renders it insoluble. The most obvious cause is removal of the water supply as a soil dries out, but other changes can also be important, such as pH, temperature, exchange sites or, in the case of elements made soluble by reduction, oxygen availability. For example, increased pH values, which often occur lower down a soil profile, may render certain iron and aluminium compounds insoluble, while a temperature decrease with depth may reduce solubility. Materials can also be redeposited if they carry an electric charge which enables them to be attracted to the exchange sites of a clay or organic colloid. Elements such as iron and manganese, which can be reduced and translocated under the anaerobic conditions associated with very wet soils, may become insoluble on moving to a drier, more aerated part of a soil, or into one in which more oxygen is available due to the oxidising effect of certain micro-organisms, or decreased mesofaunal respiration. In the case of materials carried in the form of soluble organic complexes redeposition can occur due to reduced solubility at higher pH levels, biodegradation of the organic components or polymerisation, which is also enhanced at higher pH.

Translocation in suspension will occur in the case of insoluble particles which are sufficiently light to be moved by soil water and sufficiently small to be able to pass through the soil pores. The most commonly transported materials in this context are clay and organic colloids, whose movement is aided by a lack of flocculating or cementing agents. In acid soils soluble organic matter can also enhance clay mobility by forming a hydrophilic envelope around the particles, protecting them from flocculating cations. The size and charge of colloidal particles will also influence their transport potential. For example, smectites are generally the smallest size of clay particles and are therefore most easily transported, while kaolinites are generally larger and have a relatively low charge which can be easily neutralised and are therefore amongst the least mobile types. The deposition of clays often takes the form of coatings, known as *clay skins, illuviation cutans* or *argillans,* which can be seen lining the walls of soil pores or surrounding larger mineral grains or aggregates. In soils where water movement is rapid and pores are large, material up to a few millimetres in size can be carried in suspension. This includes silt, sand, faecal pellets and other organic fragments, and in extreme conditions gravel. Deposition of this material occurs due to the evapouration of water or a reduction in pore size, and since the particles lack the electrostatic bonding properties of colloidal material, their deposits are generally confined to the upper surfaces of particles. Of course, materials transported by water are not always redeposited in the soil. Some are often lost as outputs to the ground-water drainage system and reach rivers, ultimately to be deposited in lakes or oceans.

Mechanical transfer of soil materials can occur by a variety of processes, both biological and inorganic, referred to collectively as *pedoturbation.* Of the biological processes, also known as *bioturbation,* the most important are generally soil mixing by burrowing animals and by human activity. For example, it has been estimated that earthworms can produce over 20 kg m^2 of cast material, derived usually from within a metre of the surface, while termites can bring mound-building material to the surface from depths of over 10m; erosion of the mounds can add up to an estimated 0.2 mm of soil to the surface. Earthworm burrowing can also concentrate larger material into a residual, subsurface layer or *stone line*, above which the cast material accumulates. In some instances bioturbation can produce complex layering in soils by the accumulation and burial of material moved by soil fauna. Cultivation practices, particularly ploughing, can typically mix soil to a depth of 20-50 cm, although larger implements can cause mixing to greater depths, as can excavation, such as that associated with quarrying, road construction, pipeline installation and urbanisation. Other processes may also be operative such as root decay, causing channel collapse, displacement during root growth, and root movement during strong winds, while mixing by tree-throw may also be important locally.

Inorganic mechanical transfer processes involve primarily alternating expansion and contraction of the soil as a result of wetting and drying or freezing and thawing, although freeze/thaw can also be responsible for the separation of different sizes of soil particles as well as soil mixing. Alternate wetting and drying is particularly effective where large contents of expanding lattice clay minerals are present. On drying, the clays contract and vertical cracks form in the soil, into which material can fall. When the soil becomes wet again, the clays expand, and the new infill becomes incorporated into the surrounding material. On a smaller scale, clays can become reorientated within soil aggregates by the stresses produced during expansion. This often occurs such that the clay particles become orientated preferentially with their optical axes aligned perpendicular to the direction of stress. When viewed microscopically, this gives an appearance (known as *birefringence*) similar to that of argillans formed by the redeposition of clay carried in suspension. The features are known as *pressure cutans* or *stress cutans,* and can usually be distinguished from argillans in that the latter have a sharp boundary while the boundary of stress cutans is diffuse as the reorientation gradually becomes less strong away from the point of maximum stress at which contact with other aggregates occurs.

The transfer of material resulting from freezing of the soil involves a variety of processes, some of which remain unclear. One such process is frost sorting, whereby mineral material is sorted into different sizes within the soil, either vertically or laterally. Vertical sorting can occur when soil is freezing downwards or upwards. In the former case, larger particles in the soil will allow a more rapid penetration of the freezing isotherm through them than will the surrounding soil, because of their higher thermal conductivity, therefore ice can form beneath these particles before the surrounding soil becomes completely frozen. If the particles are near the surface, the pressures created by the freezing water beneath them may be sufficient to cause upward displacement. Alternatively, if these processes are restricted to the overlying frozen soil, they may displace the unfrozen soil adjacent to the larger particles laterally or downwards. Stones can also be pushed upwards into unfrozen soil above a freezing isotherm which is advancing towards the surface. Due to lateral variations in soil thermal properties, freezing rates may vary horizontally, which can also contribute to sorting. By a combination of these processes, material can therefore become sorted into different sizes, usually with larger particles becoming concentrated towards the surface or forming various types of pattern at the surface.

Another transfer process related to soil freezing is the mixing of soil by alternate freezing and thawing, a process known as *cryoturbation*. Soil displacements have traditionally been explained in terms of *cryostatic pressure,* which results from differential freezing rates such that wet, unfrozen pockets of soil are subjected to pressures from adjacent areas where freezing is occurring. However, there is little convincing evidence to support this process—it is unlikely that unfrozen soil could be intruded under pressure into frozen material, and it is likely that unfrozen areas will be ones of water removal where pore water is therefore under tension, rather than compression. An alternative to the concept of cryostatic pressure is the cell-like movement of soil caused by different extents of ice lens formation in the initially unfrozen, or *active,* layer. It appears that soils rich in silt are most susceptible to disturbance by frost processes. These can develop large ice contents because their associated pore sizes allow water to move through the soil towards the freezing layer more effectively than larger pores associated with coarser-textured soils or the small pores of clays.

A process which can result in a similar pattern of mixing is that of density loading, where overlying, denser material sinks under gravity into underlying, less dense material, which in turn rises into the denser layer. For this to occur the soil must have an excess water supply so that intergranular contacts are lost and the soil therefore becomes liquefied. This can sometimes occur during soil melting, but can also occur in unfrozen soils if the hydrological conditions allow. The effect of density loading is to produce involutions, similar to those associated with frost disturbance. In more advanced cases, droplets of denser material become detached and sink through the underlying less dense material.

Other inorganic mixing processes include soil displacement during the growth or solution of crystals and, less commonly, by water upwelling towards the surface or during earthquakes. Mixing can also occur by differential movement of soil on a slope. Movement can occur in a variety of directions due to a combination of the processes, operating along with gravity, and at different rates at different depths beneath the surface.

Losses

Material can be lost from soils in four main forms gases, solutes, paniculate material and via vegetation removal. As in the case of additions, the processes involved can usefully be divided into surface and subsurface categories. Surface losses include gases which are produced during organic matter decomposition and lost to the atmosphere, solutes which are taken up as nutrients by vegetation and then lost when the vegetation is removed, for example by harvesting of crops or removal of trees, particulate material which is lost by water or wind erosion, and the upper parts of profiles which may be removed *en masse* by erosion or human activity. In the case of gases and solutes, the significance of losses via the surface will depend on the extent to which they are dissolved and lost by subsurface drainage, which in turn will depend on climate and land use.

Removal of particulate material by wind will be most effective in the case of soils with a high silt or fine sand content, as this size of material is more easily entrained than larger, heavier particles or small clay particles which resist entrainment due to their greater aggregation. Organic matter is also prone to wind erosion, and its lower density relative to mineral material means that larger particles can more easily be carried. Low moisture content, poor aggregation and sparse vegetation covers will also enhance susceptibility to erosion. Particle detachment occurs in response to the creation of eddies at the ground surface, and as a result of impacts from particles which are already in motion. Small particles are transported by aerial dispersion and may be carried to elevations of several thousand metres. Particles of intermediate size are transported within a metre or so of the ground surface by the process of saltation. In contrast, large particles are moved along the ground, largely as a result of impacts from saltating particles, by the process of creep. Saltation is the most important process of wind erosion in terms of the quantity of material moved, and it is estimated that 55-72 per cent of wind-eroded particles are transported in this way. In the case of erosion by water, soils with weak aggregation and low vegetation covers are particularly susceptible because the aggregates can be easily broken down by direct raindrop impact. This can also result in surface compaction and sealing to form a crust which may be several mm in thickness, which impedes infiltration, therefore enhancing the loss of material by surface runoff.

Erosion losses can also occur *en masse,* for example by glacier ice, if there is a major change in environmental conditions. Losses of material by human activity can occur in many ways, but the most common are associated with the removal of topsoil for use as a resource or during construction. For example, organic-rich soil may be removed for fuel or horticultural purposes, while mineral soil may be extracted to improve the cultivation capabilities of a soil elsewhere. Removal of soil is also often associated with the construction of roads and buildings, or with landscaping programmes.

Subsurface losses can occur in solute or solid form and can involve any of the products of addition and transformation, along with those materials undergoing transfer where conditions for total redeposition within the soil do not occur. The extent of solute out-puts will clearly depend on the solubility of the materials involved, along with factors such as temperature, which will control the rate at which reactions occur, and the speed of water movement, which will determine the time available for reactions to occur. Material lost in a solid, particulate form will only occur with any significance in soils containing large pores or other forms of passageway. In coarse-textured or uncompacted soils, pore spaces may be sufficiently large to allow material to move out of the soil and into drainage channels, but particulate losses are more usually associated with the development of soil pipes, whose diameter can range from a few centimetres up to several metres. These are found in soils which experience cracking, due to the occurrence of either highly expandable clay minerals such as smectite or high organic contents, and which also contain a subsurface layer of restricted permeability. Soil pipes allow large volumes of water to be moved rapidly through them, resulting in the removal of both solid particles and dissolved material.

SOIL PROFILES

Having examined the various processes responsible for soil formation on an individual basis, we shall now see how these processes operate in combination to produce soil profiles. In order to do so, we shall start by looking at the main characteristics of soil profiles—their formation as weathering products of the parent material and the development of horizons within them. Reference to a particular soil horizon is usually made by assigning a letter, which can then be refined by adding letters or numbers as prefixes or suffixes. A number of systems have been devised for this purpose, but two in particular have received widespread usage—the FAO-Unesco and the Soil Survey Staff schemes. Both have many characteristics in common, and indeed have many similarities with various other systems, and they will therefore form the basis of horizon designation. After examining horizon differentiation, we shall then look at the main trends or pathways along which soil formation can proceed and the principal types of profile that result.

Horizon Differentiation

The material from which a soil develops is known as the parent material; this may be bedrock or a geologically recent, superficial deposit. At the start of soil formation the parent material is unaltered. In the case of solid bedrock, this is referred to as the *R layer*. However, as weathering begins to operate, the material starts to undergo alteration and in this condition is referred to as the C *horizon*. This term is also given to unconsolidated parent material, which comprises a superficial deposit overlying bedrock, before weathering has commenced. In some instances where a superficial deposit is only thin, the weathering processes may extend down through this material into the underlying bedrock, in which case a composite C horizon may result.

With continued weathering, the soil material becomes increasingly transformed so that the original structure of the parent material is lost, and this is known as a *B horizon*, although B horizons are also recognised on the basis of transfer processes. The boundary between C and weathered B horizons is often transitional, reflecting the gradual increase in weathering from one to the other. This increase can be recognised in a number of ways. For example, there is often a decrease in the number and size of stones up through a profile as a result of an increasing intensity of weathering, and this may be accompanied by an increase in fine particles and in the amount of iron staining as iron compounds are released. The formation of clay and release of oxides during weathering may encourage aggregation and therefore allow structure to develop in the B horizon. Colour changes can also occur within a profile as iron compounds are transformed, as in the oxidation of ferric hydroxide to hematite which results in a colour change from yellow to red. Changes in texture and colour are not, however, reliable indicators of weathering if based solely on field observation, because they can also result from various transfer processes.

Progressive weathering may also change the shape and surface characteristics of mineral particles. For example, particles may become more rounded with the progression of solution weathering, or may become etched. During weathering, surface and near-surface zones of mineral particles may become altered. This can take two forms—*weathering rinds* or *grus*. Weathering rinds occur as alteration zones around a particle which remains essentially intact, for example a zone of staining resulting from iron oxidation, while grus is a zone of physical disintegration around the margin of a rock as a result of weathering, usually of igneous rocks. The thickness of weathering rinds and grus will therefore usually show an increase from the C to the B horizon.

Differences in weathering can also be seen by the use of mineral or chemical ratios. This is based on the assumption that certain minerals are broken down, or chemicals released, more easily than others, and ratios between the two will therefore vary according to the extent of weathering. For example, it is generally considered that minerals such as quartz, garnet and zircon are resistant to weathering while those such as feldspars, amphiboles and pyroxenes are less resistant. As weathering proceeds, the ratio of resistant to less resistant minerals within any particular size fraction will increase, and the magnitude of the ratio will therefore increase from lower to upper horizons. There are, however, various problems associated with the use of weathering ratios. For example, there is some uncertainty about the relative resistance of certain minerals to weathering, and also a problem concerning the statistical validity of occurrence estimates of the minerals used in ratio calculations in view of the number of grains counted. A further problem concerns the mineralogical and chemical variability of the parent material; unless this material is homogeneous, variations in mineral or chemical ratios with depth cannot be considered with any certainty to be a function solely of weathering.

A widely used alternative to weathering ratios in the observation of parent material alteration is the mineralogical investigation of clay transformation and synthesis. X-ray diffraction analysis of soil clays allows the constituent minerals to be identified and the nature of alteration can be inferred from variations in clay mineral type and proportion with depth. Soils which receive organic additions often possess a surface horizon which differs from the underlying horizons on account of its higher content of organic matter or the form in which this material occurs; this is referred to as an *O* or *A horizon*. The nature of the surface horizon will depend to a large extent on the balance between the processes of organic matter addition and its subsequent transformation, transfer and loss, and this balance will be determined by environmental conditions. In cases where the organic input greatly exceeds these other processes, there will be a large accumulation of organic matter at the soil surface. Such an accumulation is referred to in the FAO-Unesco system as an *H horizon* and is usually associated with poorly drained, anaerobic conditions in which little biological activity is occurring either in the form of organic matter decomposition or bioturbation.

When organic addition exceeds transformation to a lesser extent, but where transfer in the form of mixing remains limited, along with losses, an organic-rich surface horizon results as in the previous case, but different extents of organic matter decomposition can be recognised within it. This is referred to as an *O horizon*. Since most organic additions are via the surface, the longer material remains in the soil without bioturbation, the deeper it will become, and because organic matter decomposition will also proceed through time, the degree of decomposition will therefore increase with depth. The most recently added material, at the surface, is known as the *litter layer,* and this is usually easily recognised because much of the original structure of the material remains intact. Below this layer progressive decomposition is apparent. The original organic components are less easily identified as they become physically comminuted by the biting action of mesofauna and discoloured by the chemical attack of microorganisms. The walls of stems and roots and the veins of leaves become fragmented and the internal cellular structures begin to disappear, often becoming replaced by the faecal material of the mesofauna responsible for their ingestion. With increasing decomposition at greater depth the organic matter comprises humus and faecal material whose original components can no longer be easily recognised. O horizons are usually associated with acid soil conditions, which are unfavourable for burrowing mesofauna and therefore have a sharp boundary with the underlying mineral soil. Such conditions are also associated with an acid-tolerant flora, whose organic material is not very nutritious and is often therefore not very highly decomposed. Organic matter of this type is traditionally referred to as *mor*.

In soils where organic addition exceeds transformation, but in which transfer of organic matter in the form of mixing also occurs, although losses remain limited, the resulting surface horizon will contain a mixture of organic and mineral material. This is known as an *A horizon,* and contrasts with the previous types of horizon on account of its lower organic content and its merging boundary with the underlying material, due to mixing. Suffixes commonly applied to A horizons are *p* to denote disturbance by ploughing or other forms of tillage and, in the case of the FAO-Unesco system, *h* to denote organic accumulation where no disturbance has occurred. Although the litter layer is usually present, there is no obvious progression of increasing decomposition with depth; organic matter is derived from material which is relatively easily broken down compared to that of O horizons and therefore appears in an advanced state of decomposition at all depths, mixed with mineral material. This usually produces a crumb or granular structure. This can be made up of loosely bound aggregates containing colloidal organic and mineral material along with larger mineral and organic fragments. Soils in which mixing occurs are usually not very acid and their flora generally produce more nutritious organic matter than in the previous case. This is known traditionally as *mull*. Surface horizon organic matter whose acidity and extent of decomposition and mixing are intermediate between those of mor and mull is known as *moder*.

Soils in which the addition of organic matter is equal to or less than its transformation, transfer or loss will contain little or no organic matter in their surface horizons, since the organic material will be removed from the soil surface as quickly as it is added. It is important to stress that the precise balance between organic matter addition, transformation, transfer and loss will often result in conditions which are transitional between the categories described above. This balance will also determine the thickness of a surface horizon; in general, the greater the addition of organic matter relative to the other three processes, the thicker the horizon.

In addition to the C and B horizons produced by weathering, subsurface horizons can also form by transfer processes. These can be of two basic types those from which material is removed and those into which material is transferred. Although we have seen that there are many processes which can cause transfer, in terms of soil horizon differentiation these processes relate mainly to the transfer of material by water, rather than by mechanical means. Horizons from which material is washed are known as *eluvial* or *E horizons*, and those in which material is redeposited are termed *illuvial* or *B horizons*. In the strict sense, E horizons are defined on the basis of removal of silicate clay, iron or aluminium, either individually or in some combination. The characteristics of the horizons will obviously depend on the type of material undergoing transfer. In the case of eluvial horizons, material removed in solution will change their chemical content, and in some cases their colour and structure also. For example, if iron is removed the eluvial horizon will become lighter in colour and may also lose its structure if iron was acting as an aggregating agent. Where material is removed in suspension, a change in texture can result if mineral particles are involved. For example, the eluviation of clay and silt will cause the horizon to become coarser-textured as the fines are removed.

Illuvial horizons can be characterised on the basis of their texture, structure, colour or chemistry, depending on which soil components have been deposited, and a variety of different B horizon types can therefore be recognised. Horizons containing deposits of illuvial clay are designated *Bt horizons* and can be recognised by their finer texture compared to adjacent horizons, although this is not conclusive evidence because fine textures can also result from weathering. A more certain indication is the presence of argillans when viewed microscopically in thin section. The argillans may also give ped faces a shiny appearance, although this is not always easily visible. The presence of Bt horizons is often associated with a well developed structure, due to the aggregating effects of the illuvial clay.

Horizons which receive material in solution, for example calcium carbonate, gypsum, sesquioxides, sodium and silica may be distinguished chemically, and are given the suffixes *k, y, s, n* and *q* respectively. Some materials may also produce physical characteristics such as cementation, where accumulations are high and the soil pores become filled with deposited material, or nodules may form where localised concentrations of illuvial material accumulate. Cementation and concretion are denoted by the suffixes *m* and *c* respectively; for example, a horizon cemented by illuvial carbonates would be denoted *Bmk* while a zone of sesquioxide concretion would be denoted *Bcs*. In some cases, material carried in solution may reach the C horizon, and here the same suffixes are applied. Illuvial accumulations can also be recognised on the basis of colour as, for example, in the case of organic matter and iron which will often be darker in colour than an overlying E horizon or underlying C horizon, although again this may be the result of weathering rather than translocation in the case of iron. The deposition of these materials is seen particularly well if the soil is viewed microscopically in thin section, where the deposits often occur as coatings around mineral particles or along the walls of pores. Coatings of organic matter and iron have been termed *organans* and *ferrans* respectively. Depositional coatings of other types of illuvial material have also been given similar terms, for example *manganans* (manganese) and *silans* (silica).

Organic matter and sesquioxides are often transported in association with one another as organo-metallic complexes, and various chemical extraction procedures have been developed to identify organically bound sesquioxides. Bs horizons often possess a granular structure, although its origin is open to debate; it has been considered by some to result from the deposition of organo-metallic complexes, although others have argued that it represents residues from organic matter decomposed by soil mesofauna.

In addition to the receipt of illuvial material, B horizons can form by the residual concentration of sesquioxides after other materials have been removed. These are associated with low latitude soils and are characterised by strong weathering which results in a low silt content, a sand fraction in which no weatherable minerals remain, and a clay fraction comprising typically kaolinite and iron and aluminium oxihydrates. They also have a granular structure whose origin it is thought can relate to a combination of weathering, eluviation and bioturbation.

Soils may contain horizons which are of low porosity, known as *pans,* which restrict water movement and root growth. They can be formed by a variety of processes, and are known by a variety of terms, for example, *calcrete, plinthite, silcrete, fragipan, petrogypsic horizons* and *placic horizons*. Calcrete is a hard, indurated deposit of calcium carbonate, also known as a *petrocalcic horizon*. Carbonate deposits form initially when water droplets evapourate from the underside of stones, then gradually the pores become filled with illuvial carbonate. When this is complete, any further downward water movement is impeded, which causes water to pond up above this zone and subsequent evapouration episodes leave deposits which occur as a series of horizontal layers. Carbonate may also become concentrated by biological accumulation, or added via rainwater, atmospheric dust or sea spray. Calcrete can also form by a number of non-pedological processes, for example, the evapouration of water in shallow, ephemeral lakes or when artesian water rises to the surface.

Plinthite is an iron-rich material formed by the concentration of iron moved in solution from elsewhere, and occurs in strongly weathered soils in which clay synthesis and translocation of silica in solution also occur. On repeated wetting and drying or exposure to solar heating it hardens irreversibly to iron-stone. Silcrete, also known as *duripan*, is an indurated material cemented by various forms of secondary silica. Various origins have been proposed, including the residual concentration of silica following intense leaching of aluminium and other major cations under acidic conditions, and the leaching of silica in ground water.

Fragipans have a high bulk density and are very hard when dry, forming vertical cracks and a prismatic type of structure. They restrict water movement and are therefore often mottled due to gleying. Their origin is unclear but may relate to compaction by ground ice, with aggregation by clay minerals, iron and aluminium compounds or silica. Petrogypsic horizons are indurated layers of gypsum-rich material, the gypsum content usually exceeding 60 per cent, which again can form by translocation or by ground-water concentration. Placic horizons consist of a dark coloured pan cemented by iron, iron and manganese or an iron-organic complex, and normally do not exceed 10 mm in thickness.

Finally subsurface horizons can possess various characteristics which are not related to transfer. For example, in soils with periodic or permanent saturation, the reduction of iron resulting from the associated anaerobic conditions can produce a blue-grey colouration known as *gleying*. In zones of permanent saturation this colour occurs throughout, while in zones of periodic saturation a mottled appearance results, with anaerobic areas of reduced iron being interspersed with orange-brown, aerobic areas of oxidised iron. Variations in the degree of aeration occur because of different rates of wetting and drying due to small-scale variations in soil hydrological properties, particularly porosity. Gleyed horizons are given the suffix g, which can be applied to both B and C horizons, for example Btg and Cg.

In some cases, additions to a soil profile may be so great that the material does not form a horizon by the processes described above, but buries any existing horizons that have already developed. This can occur, for example, if a soil is inundated by a river, the sea or an ice sheet, which covers the soil with thick deposits of sediment. A similar situation could occur in the case of volcanic activity, mass movement, aeolian transport or human activity. Any type of horizon can become buried in this way, and is denoted by the suffix *b*. Buried horizons can be of great importance in the study of environmental history.

Mechanical transfer processes can also alter profiles, for example in the case of cryoturbation which can cause horizon boundaries to become contorted or, in more extreme cases, portions of horizons may become detached and incorporated into adjacent horizons. Disturbance by bioturbation or downslope movement can inhibit horizon development, causing the soil to become homogenised. Soil losses can cause disturbance either to individual horizons or to complete profiles. For example, a surface mineral-organic horizon may become partially or completely removed by excavation, while erosion may result in the loss of both A and B horizons. Profiles truncated in this way can, like buried soils, be useful in examining environmental history.

Pedogenic Pathways

The combination of horizons within a soil forms a soil profile. Profiles will vary according to the nature of the horizons they contain, which will be determined by the way in which the soil-forming processes operate, and this in turn is influenced by environmental conditions. Because these conditions vary widely over the earth's surface, an almost infinite variety of soil profiles results. However, it is possible to recognise a number of basic pathways along which soil formation can proceed, which results in a number of basic categories of soil type.

In environments where weathering is not excessively restricted by environmental factors, a weathered B horizon will develop above the C horizon, and this can be recognised in terms of various physical, chemical and mineralogical characteristics. The most strongly weathered profiles have been categorised into three types —*fersiallitic, ferruginous* and *ferrallitic*. Fersiallitic soils are characterised by clay mineral synthesis, and the dehydration of iron oxides and subsequent crystallisation to hematite, a process known as *rubification*. This gives the profile a reddened appearance, as in the case of reddened soils occurring over limestone, known as *terra fusca* where rubification is slight or *terra rossa* where it is more marked. Ferruginous soils show greater clay synthesis and also the partial solution of silica, although rubification may not occur. Ferrallitic soils, which are also often rubified, represent the most advanced stage of weathering, with greater silica mobility and the formation of kaolinite.

Many soils, however, can be grouped according to the nature of transfer processes. For example, in dry environments where there is a net upward movement of water, alkaline components, particularly salts, which would normally be easily leached downwards or out of the profile, are moved upwards towards the surface by capillary action, where they can accumulate as a thin surface crust or as a horizon. Where salts are involved, the resulting profiles are known as *solonchak*. They are light coloured, and usually have a low organic content and alkaline pH values. Two soil types related to solonchak are the *solonetz* and *solod*. These result from leaching, which initially causes dispersion of the organic matter and hydrolysis of salts, producing a very high pH; this is the solonetz phase. With continued leaching the organic matter is moved further down the profile and the pH in the upper part of the profile decreases as the salts are also leached downwards, giving a solod. Where calcium is the major element involved in transfer, either by leaching or upward capillary movement, the process is known as *calcification,* and in extreme cases can lead to the formation of calcrete. In cases of more moderate calcification, *chernozems* can form; these are associated with grassland areas and have a mull surface horizon and neutral or slightly alkaline pH values.

If the extent to which alkaline components are leached exceeds the rate at which they are replaced by weathering, the soil will gradually become more acid in its upper part, as the exchange sites on the clay-humus complex become occupied by hydrogen ions and the alkaline components, particularly sodium and potassium, are lost from the profile via ground drainage waters. Soils which show a limited degree of acidification are often associated with broadleaf forest and are known simply as *brown earths.* They have a mull or moder surface horizon, and in the more alkaline varieties of profile, the main transfer processes are the mixing of organic matter by burrowing organisms, and only slight leaching of basic cations. In the more acid variety, leaching of basic cations can be greater, resulting in less aggregation in the upper part of the profile. This can allow clay to become washed downwards, a process known as *lessivage,* which may therefore form a Bt horizon showing characteristic argillans. These profiles are normally slightly to moderately acid, with mull or moder types of organic matter and pH values in the order of 4.5-6.5. Such profiles are sometimes referred to as *argillic brown earths,* the term argillic being used to distinguish these types from ordinary brown earths in which clay movement is absent or limited. Although these are characterised by the presence of clay translocation, the amount of illuvial clay in the Bt horizon is often much lower than that of the clay produced by *in situ* weathering. Clay translocation and Bt horizon formation also occurs in weathered, fersiallitic soils, but decreases in ferruginous and ferrallitic soils because the kaolinite which is formed is resistant to movement by water.

Soils in which sesquioxides are transferred are usually moderately to highly acid, either because they are developed from an acidic parent material or because most of the alkaline components have already been leached out of the profile. Surface horizons are generally of the mor type, with pH values below 4.5. The movement of sesquioxides is traditionally associated with organic complexing, which is enhanced under acidic conditions, and the resulting profiles develop a lightcoloured E horizon from which the iron staining has been removed, overlying a zone of illuviation. These profiles are known as *podzols* and can be divided into two main types, those in which the illuvial zone comprises a single horizon of the Bs type, known as *ferric* podzols, and those which have an additional organic-rich illuvial horizon between the E and Bs, known as *humo-ferric* podzols. The formation of Bs horizons in some cases the illuvial zone may contain a layer of strong iron cementation, often associated with restricted drainage, which is thought to form where illuviation is strong in relation to bioturbation. The importance of the organic complexing process in the formation of podzols has become questioned; it has been argued that sesquioxides are concentrated in B horizons largely by inorganic processes involving translocation in solution, with the additional organic complexing with aluminium occurring within the B horizon following the attack of the clay minerals imogolite and proto-imogolite. It may, however, be the case that the extent to which organic and inorganic processes are involved in podzol formation is determined by the amount of organic matter present, with organic complexing being more important in soils with a higher organic matter content.

Highly leached and weathered profiles are found mainly in low latitude regions and are known traditionally as *laterites,* being characterised by the process of ferrallitisation. Their horizon characteristics vary greatly, depending on the materials undergoing transfer, but two basic sets of processes can be distinguished. The first type, referred to as *latosolisation,* is associated with mafic rocks high in weatherable minerals, for example basalt, in which silica and bases are leached out of the upper part of the profile, leaving a residual accumulation of sesquioxides. The second type, known as *lateritisation,* involves the formation of plinthite, which, because the concentration of the iron is often not even, commonly takes a nodular form. Laterite profiles are often much deeper than the other types discussed above due to intense weathering operating over long periods of time, with depths of 100 m or more being recorded; profiles of the other soil weathered, fersiallitic soils, but decreases in ferruginous and ferrallitic soils because the kaolinite which is formed is resistant to movement by water.

Environmental conditions can cause soils to be poorly developed. In such profiles all the pedogenic processes may be restricted, and some may be absent entirely. Although a soil, by definition, must show evidence of some weathering or organic matter accumulation, it does not necessarily need to experience significant transformations or transfers of material, and a poorly developed profile can therefore possess only a weathered C horizon and/or a surface organic horizon, with no B horizon. Soils which contain only an A and C horizon are often referred to simply as A/C profiles. They may also be known as *lithosols* if developed from weakly weathered bedrock, *regosols* if formed in regolith, or *rankers* if occupying steep slopes. A special category of poorly developed profile occurs in the case of soils developed on limestone parent material, where the absence of a B horizon does not necessarily result from limited weathering but from the fact that all the weathering products are removed from the profile in solution; this type of soil is known as a *rendzina.* Another special category of soil is the *andosol* or *andisol* which develops from parent materials of airborne volcanic deposits and is distinguished from other soils by its low bulk density and high content of amorphous or low crystallinity weathering products.

The profiles described above form in freely draining conditions, but there are many soils whose profiles develop as a result of restricted drainage. These can be divided into organic and mineral varieties. Where transformations, transfers and losses are restricted by poor drainage, organic additions will undergo little decomposition, resulting in the formation of a *peat;* this process is sometimes known as *paludification*. Peats may be of varying thickness and pH, but all usually show the water table close to the surface. Within mineral soils, restricted drainage causes gleying and in cases of permanent saturation, transfer processes are also inhibited due mainly to lack of water movement and bioturbation. Profiles of this type are known as *gleys*, although gleying can also occur within other soil types in which saturation is periodic. Two basic types of gley can be recognised—where saturation occurs due to the local water table extending up into the profile, gleying increases with depth and these profiles are termed *ground-water gleys*. In contrast, profiles which have a perched water table in their upper part, caused for example by compaction or clay illuviation, and become progressively well drained with depth are called *surface-water gleys*.

It is important to note that the pedogenic pathways outlined above relate to processes which characterise particular types of soil, but that these may not be the only processes operating in these soils. For example, clay translocation can operate in rubified soils and gleying can occur in brown earths and podzols if the drainage is impeded.

Soil Classification

The profiles discussed above allow a basic distinction between soil types, but this is obviously generalised and qualitative. There have, however, been numerous attempts throughout the history of soil science to classify soil profiles in a more detailed and systematic way. Soils may be classified for a variety of reasons, but here we will consider only genetic classification systems. Most of these use a hierarchical approach, whereby a series of major groups are identified and then each is subdivided at a number of levels of increasing detail.

As in the case of horizon designation, probably the best known and most widely used systems of classification are those developed by the FAO-Unesco and Soil Survey Staff, and these will be focused on here. The FAO-Unesco system, based on an earlier scheme, uses 28 *major soil groupings*, which are in total subdivided into 153 *units*. The names of the units are derived from a variety of linguistic sources, some of the terms simply being adopted from existing, traditional terms and others being newly devised. Many of the characteristics used for classification are morphological, although occasionally refer to processes or chemistry. Division of the units ranges from two in the case of Greyzems, to nine in the case of Cambisols. These divisions are made according to terms derived mainly from Greek and Latin roots, most of the terms being morphological, and some being the same as those used to identify the soil units themselves. Such a system, while being fairly comprehensive, is only qualitative, so in order to make it less subjective, various sets of quantitatively defined diagnostic horizons and properties have been devised for allocating soils to particular units and sub-units. Five types of surface horizon and five types of B horizon are recognised, along with six other horizon types, and 26 diagnostic properties are defined, mainly on the basis of morphological and chemical properties.

The groups are known as *orders* and, as in the case of the FAO-Unesco system, their names are derived from a variety of linguistic roots. Some refer to morphological characteristics, while others relate to processes or chemical properties. The orders are divided into 47 *sub-orders* using terms again derived mainly from Greek and Latin sources and based on a variety of characteristics. The number of sub-orders ranges from two to seven per order, with most having four or five. Again, a number of diagnostic horizons and properties are also recognised. The sub-orders are divided into *great groups* using a variety of terms based on the diagnostic properties, which gives around 225 categories. Further division is made into *sub-groups* by adding various adjectives to the great group names, and still further subdivision can be made into *families*, primarily on the basis of particle size, mineralogy and temperature regime. A final level of subdivision, the *series*, is also possible, but this is usually derived from the name of the locality in which that type of soil was first recognised, and therefore has no real value in providing information about soil characteristics or genesis; series are used in soil mapping. Although this scheme provides the most comprehensive soil classification available, it recognises that not all soils will be accommodated within it, and additional categories have been proposed, particularly for soils influenced by human activity such as waste disposal, construction and dredging.

Hierarchical classification systems have also been developed elsewhere for use within a national context. The number of possible designations is less than in the Soil Survey Staff system, but still allows a fairly comprehensive classification. For example, the British Isles system has three levels within the hierarchy, comprising six major soil groups, 23 soil groups and 94 sub-groups.

Despite the widespread use of hierarchical classification systems, various problems are associated with them. An obvious problem for many non-specialist users is the complex and unfamiliar terminology, which makes the classification and recognition of soil types rather laborious. Another problem is that the criteria used to distinguish soils at any one level within a hierarchy are not all defined using the same characteristics; for example, a mixture of morphological, genetic and chemical characteristics is used in defining the first level of the hierarchy in the systems discussed above. The use of quantitatively defined criteria for differentiation of soils means that certain soil types which are very similar in many respects can be allocated to different categories if they narrowly fail to meet the criteria being used. This occurs, for example, in the case of the Spodosol category of the Soil Survey Staff system, where podzols cannot be classified as such because their B horizons do not comply with the criteria used for defining a spodic horizon. The opposite situation could also occur in the case of soils which are in many respects different being placed in the same category if they happen to meet the differentiating criteria which have been selected. A further problem is the difference in the number of subdivisions of the various categories at any one level within the hierarchy. For example, using the FAO-Unesco system, Cambisols can be allocated to one of nine possible categories whereas only two subdivisions of Greyzems are possible. Finally, it is clear that selection of the differentiating criteria is subjective, with only a limited number of numerous possible soil characteristics being chosen and value judgements being made about the relative importance of criteria both within and between hierarchical levels.

Some of the problems associated with traditional classification systems have been overcome by the development of numerical methods. A variety of methods are available, but the two most commonly used are ordination and the construction of dendrograms. Ordination methods involve the use of principal-components analysis and can be considered as follows. If the relationships between a range of soil properties for a number of soils are considered to be represented by a series of points in a multidimensional space, these can be reduced mathematically to one or a few principal axes which account for as much of the total variance as possible and in this form the differentiation between the soils can become relatively easy. Projection of the sites on to the plane defined by the first two principal axes therefore gave the most informative single display of relationships between the soils. Rotation of the factor axes allows each of the original variates to contribute strongly to one of the factors and much less so to the others, making the plot more simple to interpret. As a result, the first axis strongly represented properties related to the moisture of the soil, while the second axis related strongly to textural properties. The soils are therefore grouped in Figure such that the driest soils are on the right and the wettest on the left, while the heaviest-textured soils are at the top and the lightest at the bottom. The small number of sites located in the bottom lefthand quarter of the plot therefore indicates that there are relatively few wet, light-textured soils.

The construction of dendrograms is concerned with the distances between the sites plotted as a result of principal-components analysis, and is an alternative hierarchical approach to the systems discussed earlier. The most closely grouped sites form the lowest level of the hierarchy and increasingly distant groups are linked at higher levels. However, although hierarchical numerical classifications are often applied to soils, they are designed for populations which show nested clustering, which is rarely found in the case of soils; consequently non-hierarchical numerical methods such as canonical variate analysis may be used as an alternative.

The advantage of numerical methods is that they allow a large number of soil characteristics to be taken into account without any preconception of their relative importance. They do, however, suffer from the problem that no single attribute is either sufficient or necessary to confer class membership, and it is therefore very difficult to construct an identification key. They also require a large quantity of data for each soil being classified, and to date their use has therefore been confined to the local, rather than national, scale.

Chapter 3

Properties of Soil

SOIL

Soil may be defined as a thin layer of earth's crust which serves as a natural medium for growth of plants.

Soil Structure

It refers to the arrangement of soil particles. It is one of the important property of soil, since it influences aeration, permeability and water capacity.

Types of Structure

- Platy - Horizontal alignment
- Prism like - Columnar type
- Block like - Angular or sub- angular types
- Spiroidal - Granular and crumb types

Soil Texture

- The varying proportions of particles of different size groups in a soil constitute is known as soil texture.
- The principle textural classes are clay, clay loam, sandy clay, silt clay, sandy clay loam, silty clay loam, sandy loam, silt loam, sand, loamy sand and silt.

Soil Profile

- It is the vertical section of the soil through all its horizons from the surface to the unaffected parent materials". Generally the profile consists of three mineral horizons viz., A, B and C.
- The surface soil or that layer of soil at the top which is liable to leaching and from which some soil constituents have been removed is known as horizon 'A' or the horizon of eluviation. The intermediate layer in which the materials leached from horizon 'A' have been re-deposited is known as horizon 'B' or the horizon of illuviation. The parent material from which the soil is formed is known as horizon ' C'.
- The soil in each of these horizons is usually uniformly developed and presents a more or less homogeneous character. Each layer or horizon develops specific morphological features such as the size and shape of particles, their arrangement, colour, consistence etc. which distinguish from one horizon to another.
- Study of soil profile is important since it reveals the characteristics and qualities of the soil.

Soil Composition

- Organic matter
- Soil organisms - Micro flora and Micro fauna.
- Soil water
- Soil air
- Inorganic matter - Macro nutrients and Micro nutrients

Organic Matter

- The plants and animals grown in weathered material and the organic residues left behind decay with time and become an integral part of the soil. The main source of soil organic matter is plant tissue. Animals are subsidiary source of soil organic matter.

- The micro flora like bacteria, fungi, algae, actinomycetes, and micro fauna like protozoa, nematodes, macro fauna like earthworms, ants etc. play an important role in formation of organic matter.
- The organic matter influences the soil in respect to colour, physical properties, supply of available nutrients and adsorptive capacity.

Soil Organisms

- Soil is the habitat for enormous number of living organisms. Some of these organisms are visible to naked eye where as others can be seen by microscope only.
- Roots of higher plants are considered as soil macro flora while bacteria, fungi, algae and actinomycetes are considered as soil micro flora. Protozoa and nematodes are the significant soil micro fauna where as the earthworms, moles and ants constitutes soil macro fauna.

Soil Water

- In order to function as a medium for plant growth, soil must contain some water. The main functions of water in the soil are as follows:
- Promotes many physical and biological activities of soil.
- Acts as a solvent and carrier of nutrients.
- As a nutrient itself.
- Acts as an agent in photosynthesis process.
- Maintains turgidity of plants.
- Acts as an agent in weathering of rocks and minerals.

Soil Air

- Oxygen is essential for all biological reactions occurring in soil. Its requirement is met from the soil air.

- The gaseous phase of soil acts as a path way for intake of oxygen which is absorbed by soil micro organisms, plant roots and for escape of carbondioxide produced by the plants.
- This two way process is called soil aeration. Soil aeration become critical for the plant growth when water content is high, because water replaces soil air.

Soil Inorganic Matter

- The inorganic constituents of the soil comprises carbonates, soluble salts, free oxides of iron, aluminium and silica in addition to some amorphous silicates.
- The inorganic constituents forms the bulk of the solid phase of the soil. Soils having more than 20per cent of the organic constituents are designated as organic soils.
- Soils where inorganic constituents dominates they are called mineral soils. The majority of the soils in India are mineral soils.

Soil pH

The negative logarithm of hydrogen ion (H +) concentration is called pH. Soil pH may be acidic, basic or neutral.

Soil Fertility

Soil fertility deals with the nutrient status or ability of soil to supply nutrients for plant growth under favourable environmental conditions such as light, temperature and physical conditions of soil.

Soil Productivity

Soil productivity is defined as the capability of the soil for producing a specified quantity of plant produce per unit area and the ability to produce sequence of crops under a specified system of management.

PROBLEM SOILS

The soils which owe characteristics that they can not be economically used for the cultivation of crops without adopting proper reclamation measures are known as problem soils.

Acid Soils

Those soils with pH less than 6.5 and which respond to liming may be considered as acid soils.

Reasons for Acidity

- Humus decomposition results in release of large amounts of acids. There by lowering the pH.
- *Rainfall*: In areas with more than 100 cm rainfall associated with high R.H., Ca, Mg is dissolved in water and leached out due to this base saturation of soil decreases.
- Application of elemental sulphur under goes reactions resulting in formation of H_2Sc_4.
- Continuous application of acid forming fertilizers like ammonium sulphates or ammonium chlorides results in depletion of Ca by CEC (cation exchange capacity) phenomenon.
- *Parent Material*: Generally rocks are considered as acidic, which contain large amount of silica (Sio_2) when this combined with water, acidity increases.

Characteristics

- PH is less than 6.5
- This soils are open textured with high massive Structure.
- Low in Ca, Mg with negligible amount of soluble salts.

- This soils appear as brown or reddish brown, sandy loams or sands.

Injury to Crops

- Plant root system does not grow normally due to toxic hydrogen ions.
- Permeability of plant membranes are adversely affected due to soil acidity.
- Enzyme actions may be altered, since they are sensitive to PH changes.

Indirect Affects

- Deficiency of Ca and Mg occur by leaching.
- Al, Mn and Fe available in toxic amounts.
- All the micro nutrients except molybdenum are available. So 'Mo' deficiency has been identified in leguminous crops.
- Phosphorous gets immobilized and its availability is reduced.

ACTVITY OF MICRO ORGANISMS

Most of the activities of beneficial organisms like Azatobacter and nodule forming bacteria of legumes are adversely effected as acidity increases.

Amelioration

- Lime as reclaiming agent: Lime is added to neutralize acidity and to increase the PH, so that the availability of nutrients will be increased.
- Basic slag obtained from Iron and steel industry can be substituted for lime. It contains about 48-54per cent of CaO and 3-4per cent MgO.
- Ammonium sulphate and Ammonium chloride should not be applied to acid soils but urea can be applied.
- Calcium Ammonium Nitrate (CAN) is suitable to acid soils.

- Any citrate soluble phosphate fertilizer is good source of phosphorous for acid soils.
- Eg. Dicalcium phosphate (DCP), Tricalcium phosphate (TCP)
- Potassium sulphate is a suitable source of 'K' for acid soils. But MOP is better than K_2So_4 because Cl of MOP replaces –OH ions, their by release of -OH ions tends to increase the PH.

Alkaline Soils

Alkali soils are formed due to concentration of exchangeable sodium and high pH. Because of high alkalinity resulting from sodium carbonate the surface soil is discoloured to black; hence the term black alkali is used.

Reasons for Alkalinity

- The excessive irrigation of uplands containing Na salts results in the accumulation of salts in the valleys.
- In arid and semi arid areas salt formed during weathering are not fully leached.
- In coastal areas if the soil contains carbonates the ingression of sea water leads to the formation of alkali soils due to formation of sodium carbonates.
- Irrigated soils with poor drainage.

INJURY TO CROPS

- High exchangeable sodium decreases the availability of calcium, magnesium to plants.
- Dispersion of soil particles due to high exchangeable 'Na' leads to poor physical condition of soil, low permeability to water and air, tends to be sticky when wet and becomes hard on drying.
- Toxicity due to excess hydroxyl and carbonate ions.
- Growth of plant gets affected mainly due to nutritional imbalance.

- Restricted root system and delay in flowering in sensitive varieties.
- Typical leaf burn in annuals and woody plants due to excess of chloride and sodium.
- Bronzing of leaves in citrus.
- It effects the solubility of zinc(Zn).

CROPS SUITABLE FOR CULTIVATION IN ALKALINE SOILS

Barley, Sugarbeet, Cotton, Sugarcane, Mustard, Rice, Maize, Redgram, Greengram, Sunflower, Linseed, Sesame, Bajra, Sorghum, Tomato, Cabbage, Cauliflower, Cucumber, Pumpkin, Bitterguard. Beetroot, Guava, Asparagus, Banana, Spinach, Coconut, Grape, Datepalm, Pomegranate.

Amelioration

- The process of amelioration consists of two steps.
  - To convert exchangeable sodium into water soluble form.
  - To leach out the soluble sodium from the field. Amendments used for reclamation of Alkali soils.

Gypsum

- It is slightly soluble in water. So it should be applied well in advance.

Requrement

- For every 1 m.e of exchangeable Na per 100 gm of soil, 1.7 tonns of Gypsum/ acre is to be added.

Application

- If the requirement is 3 tonnes/ acre- apply in one dose.
- If the requirement is 3 to 5 tonnes/acre- apply in 2 split doses.
- If the requirement is 5 or more tonnes/ acre - apply

in 3 split doses.

- Use of Pyrites (FeS_2)
- Sulphur present in pyrites causes decrease in pH of soil due to formation of H_2So_4.
- $H_2So_4 + CaCo_3 \rightarrow Ca\ S0_4CaSo_4 + Na \rightarrow Na\ So_4 + Ca$ (leachable)
- Application of sulphur.
- Application of molasses.
- Drainage channels must be arranged around the field.
- Growing the green manure crops and incorporate in the field.

ParametersDetailspHmore than 8.3ECLess than 4 m.mhos/ cmESPMore than 15Chemistry of soil solutionDominated by carbonate and bicarbonate ions and high exchangeable sodium.Effect of electrolyte on soil particles Dispersion due to high amount of exchangeable sodiumAdverse effect on Plant Alkalinity of soil solutionGeographic distributionSemi arid and semi humid - areas.Diagnosis under field condition Presence of dispersed soil surface. Columnar structures present in the sub-soil

Saline Soils

The saline soils contains toxic concentration of soluble salts in the root zone. Soluble salts consists of chlorides and sulphates of sodium, calcium, magnesium. Because of the white encrustation formed due to salts, the saline soils are also called white alkali soils.

Reasons For Salinity

- In arid and semi arid areas salts formed during weathering are not fully leached. During the periods of higher rainfall the soluble salts are leached from the more permeable high laying areas to low laying areas and where ever the drainage is restricted, salts accumulate on the soil surface, as water evaporates

- The excessive irrigation of uplands containing salts results in the accumulation of salts in the valleys.
- In areas having salt layer at lower depths in the profile, seasonal irrigation may favour the upward movement of salts.
- Salinity is also caused if the soils are irrigated with saline water.
- In coastal areas the ingress of sea water induces salinity in the soil.

Parameters	Details
PH	Less than 8.3
Ec	More than 4.0 m.mhos/ cm
ESP (exchangeable sodium per cent)	Less than 15
Chemistry of soil solution	Dominated by sulphate and chloride ions and low in exchangeable sodium
Effect of electrolytes on soil particles	Flocculation due to excess soluble salts.
Main effect on plant	High osmotic pressure of soil solution
Geographic distribution	Arid and semi arid regions.
Diagnosis under field condition	Presence of white crust Presence of chloris barborata(weed)Patchy growth of plants.

Injury to Crops

- High osmotic pressure decreases the water availability to plants hence retardation of growth rate.
- As a result of retarded growth rate, leaves and stems of affected plants are stunted.
- Development of thicker layer of surface wax imparts bluish green tinge on leaves

- Due to high EC germination per cent of seeds is reduced.

Crops Suitable For Cultivation In Saline Soils

Barley, Sugarbeet, Cotton, Sugarcane, Mustard, Rice, Maize, Redgram, Greengram, Sunflower, Linseed, Sesame, Bajra, Sorghum, Tomato, Cabbage, Cauliflower, Cucumber, Pumpkin, Bitterguard. Beetroot, Guava, Asparagus, Banana, Spinach, Coconut, Grape, Datepalm, Pomegranate.

Amelioration

- The salts are to be leached below the root zone and not allowed to come up. However this practice is some what difficult in deep and fine textured soils containing more salts in the lower layers. Under this conditions a provision of some kind of sub-surface drains becomes important.
- The required area is to be made into smaller plots and each plot should be bounded to hold irrigation water.
- Separate irrigation and drainage channels are to be provided for each plot.
- Plots are to be flooded with good quality water upto 15 - 20 cms and puddled. Thus, soluble salts will be dissolved in the water.
- The excess water with dissolved salts is to be removed into the drainage channels.
- Flooding and drainage are to be repeated 5 or 6 times till the soluble salts are leached from the soil to a safer limit.
- Green manure crops like Daincha can be grown upto flowering stage and incorporated into the soil. Paddy straw can also be used.
- Super phosphate, Ammonium sulphate or Urea can be applied in the last puddle. MOP and Ammonium chlorides should not be used.

- Scrape the salt layer on the surface of the soil with spade.
- Grow salt tolerant crops like sugar beet, tomato, beet root, barley etc
- Before sowing, the seeds are to be treated by soaking the seeds in 0.1per cent salt solution for 2 to 3 hours.

Soil Testing

- *Need*: When land is brought under cropping, grain or fruit and sometimes the entire plants are removed (harvested) from the land. Hence, the soil losses a considerable amount of its nutrients (up take by plants). If cropping is continued over a period of time, without nutrients being restored to the soil, its fertility will be reduced and crop yields will decline.
- Apart from removal by crops, nutrients may also be lost from the soil through leaching and erosion. Even to maintain soil productivity at the existing levels, it is necessary to restore to the soil, the nutrients removed by crops as also those lost through leaching and erosion.
- Continued maintenance of a high level of soil fertility is an indispensable for profitable land use and sustained agricultural production. From time to time the inherent fertility of soil has to be evaluated. There are different methods for soil fertility evaluation as listed below:

Visual Method of Diagnosis

- Plant analysis (Analysis of whole or part of plant growing on the soil in question).
- Biological tests in which higher plants or certain micro organisms are used.
- Soil tests.
- Field experiments.

Advantages

- Among the different methods, soil testing is a better method for the following reasons.
- Soil testing, being a rapid method, is an advantage over the biological methods which are relatively elaborate and time consuming. It is also better than deficiency symptoms and plant and tissue analysis, because the needs can be determined before the crop is planted while in the other methods the crop needs can be ascertained only after the crop is grown, by which time it may be late to correct any nutritional deficiencies that may be indicated.
- The main purpose of soil testing is to evaluate the fertility status of the soil. It provides a basis for fertilizer, lime and gypsum recommendation. Laboratory test is a means of making an inventory of the chemical conditions of soil and determining treatments, if any, are needed. Soil test information is then used along with an evaluation of specific crop requirements, cropping history and physical characteristics of the soil for determining the exact amounts of different nutrients and soil amendments, if any, needed for a certain crop or cropping sequence. With this objective in view, a number of soil testing laboratories have been established in the country by the State Governments, Agricultural Universities and fertilizer industry for making fertilizer recommendations to farmers on the basis of the fertility status of their soils. This service is generally rendered free of cost.

Methodology

Soil Sampling: Soil tests and their interpretation are based on the soil samples sent in for analysis. It is, therefore, important that soil sample should be properly collected and be representative of the area to be tested. Soil tests and their interpretation are as reliable as the soil samples drawn.

Sampling Procedure

- Each field should be sampled separately. When the areas within the field distinctly differ in crop growth, in the appearance of soil, in elevation or area known to have been manured or cropped differently in the field should be divided suitably and each area sampled separately.
- Drawing samples from spots which do not represent the field should be avoided. Such spots may be old bunds, marshy spots, hedges, areas previously occupied by compost heaps, etc. Sampling should not be done in a field within three months of the application of lime, ash or fertilizer.
- Proper sampling tool should be used. Samples can be satisfactorily taken with a soil tube, an auger, a kassi (spade) or khurpi. In a very friable soil, a large spoon can also be used.
- A composite sample may be taken from each area. After scrapping the surface litter, a uniform core or a thin slice of soil from the surface to plough depth (15 to 22 cm deep) from 15 to 20 spots should be taken. In a hard soil, a small pit of about 15 cm × 15 cm and of about 15 cm in depth be made. Than a v-shaped slice from one of the slides be taken.
- Where crops have been planted in lines, sampling may be done between the lines.
- Individual cores or slices should be collected in a clean container. All lumps should be broken and mixed well in the container or on a clean cloth. The size of the composite sample should be reduced by successive quartering to about half a kilogram.
- The sample should be dried in shade for an hour or two before putting it into a bag and dispatching it to the nearest soil testing laboratory. Alkathene bags which are available from soil testing laboratories or ordinary clean cloth bags may be used.

- Each sample should be identified by name or number to correspond to the field name or number and also by the cultivator's name.
- The information sheet furnished by the soil testing laboratory should be filled up completely. This is important and will help the chemist to schedule a more accurate fertilizer recommendation. The information sheet along with the soil sample container should be sent to the soil testing laboratory.
- If standard information sheets are not available, information may be given on the following points.

SOIL TEST INTERPRETATION AND FERTILIZER RECOMMENDATIONS

- From the results of analysis of soil samples sent by the farmer and information sheet supplied by him, soil test reports are prepared in the laboratories. Copies of these reports are sent to the concerned farmer.
- Soil test reports are usually in three main parts. First part indicates results of analyses of the soil sample. Most laboratories give actual analyses as well as the ratings. Second part is fertilizer recommendations for the crop based on soil analyses, history of the field like cropping pattern, manures and fertilizers earlier applied, etc. This part indicates quantities of nitrogen (N), Phosphate (P_2O_5), Potash(K_2P), Zinc (Where facilities exist) and also of lime or gypsum to be applied per hectare. Most laboratories also show in the report optimum quantities of organic manures as per recommendations of the Agriculture departments.
- The third part of the report usually indicates time and methods of fertilizer application and other practices required to make the fertilizer use more efficient.

During the relatively short period that soil testing service has been in operation in this country a large number of soil samples have been analyzed in various laboratories. Based on the results of these analyses, soil fertility maps have been prepared indicating the nutrient status of nitrogen, phosphorus, potassium and zinc in different parts of the country. It must however, be noted that this is only a broad classification, since it is based on limited soil sample analysis.

INFRASTRUCTURE RELATED INFORMATION

The basic infra structure needed for soil testing are PH meter, conductivity meter, spectrophotometer or calorimeter for estimating phosphorous, flame photometer for potassium and atomic absorption spectrophotometer for estimating micro nutrients. Besides the instruments, glass - ware and necessary chemicals are required.

- In most of the soil testing laboratories PH, Ec, Organic carbon (as an index of available nitrogen), available phosphorus, available potassium are estimated. If necessary micro nutrients like Fe, Cu, Mn, Zn and B are estimated, Ca, Mg and S are also estimated if any deficiency symptoms are observed on crops.
- PH is estimated by a glass electrode PH meter in 1:2 soil water suspension.
- Electrical conductivity is measured by a conductivity meter in 1:2 soil water suspension.
- Available nitrogen is estimated by Subbaiah and Asija (1956) method (distillation of soil with alkaline potassium permanganate solution). But in most of the laboratories organic carbon is taken as an index of available nitrogen content of soil assuming C:N ratio is 10. Organic carbon is determined by chromic acid oxidation by rapid titration (Walkley and Black (1934) rapid titration).

- Phosphorous is determined by Olsen's (1954) using 0.5 M sodium bicarbonate as extractant and phosphorus is analyzed calorimetrically.
- Neutral normal ammonium acetate solution is the most widely used extractant for available potassium which is analyzed by flame photometer.
- Micro nutrients are extracted by DTPA and determined by atomic absorption spectro photo meter.

CLASSIFICATION OF INDIAN SOILS

There are 8 major group of soils in India which are furnished below

Red Soils

- Red colour is due to various oxides of iron. They are poor in N, P, K and with pH varying 7 to 7.5. These soils are light textured with porous structure. Lime is absent with low soluble salts.
- Red soils occurs extensively in Andhra Pradesh, Assam, Bihar, Goa, Parts of kerala, Maharastra, Karnataka, Tamilnadu and West Bengal. Most of the red soils have been classified in the order ' Alfisols'.

Lateritic Soils

- Seen in high rainfall areas, under high rainfall conditions silica is released and leached down wards and the upper horizons of soils become rich in oxides of iron and alluminium. The texture is light with free drainage structure.
- Clay is predominant and lime is deficient. pH 5 to 6 with low in base exchange capacity, contained more humus and are well drained. They are distributed in summits of hills of Daccan karnataka, Kerala, Madhyapradesh, Ghat regions of Orissa, Andhra pradesh, Maharastra and also in West Bengal, Tamilnadu and Assam.

- Most of the laterite soils have bee classified in the order ' ultisols' and a few under ' oxisols'.

Alluvial Soils

- These are the most important soils from the agriculture point of view. The soils are sandy loam to clay loam with light grey colour to dark colour, structure is loose and more fertile. But the soils are low in NPK and humus.
- They are well supplied with lime; base exchange capacity is low, pH ranges from 7 to 8. These soils are distributed in Indo-Gangetic plains, Brahmaputra valley and all most all states of North and South. Most of the alluvial soils have been classified in the orders ' Entisols', ' Inceptisols' and ' Alfisols'.

Black Soils

- This is well known group of soils characterised by dark grey to black colour with high clay content.
- They are neutral to slightly alkaline in reaction. Deep cracks develop during summer, the depth of the soil varies from less than a meter to several meters. Poor free drainage results in the soils, base exchange is high with high pH and rich in lime and potash. Major black soils are found in Maharastra, Madhyapradesh, Gujarat and Tamilnadu.
- Cotton is most favourable crop to be grown in these soils. These soils are classified in the order 'Entisols', ' Inceptisols' and ' vertisols'.

Forest Soils

- This group of soils occur in Himalayas. Soils are dark brown with more sub-soil humus content. They are more acidic.

Desert Soils

- These soils are mostly sandy to loamy fine sand with brown to yellow brown colour, contains large amounts of soluble salts and lime with pH ranging 8.0 to 8.5. Nitrogen content is very low.

- The presence of Phosphate and Nitrate make the desert soils fertile and productive under water supply. They are distributed in Haryana, Punjab, Rajasthan. They are classified in the order ' Aridisols' and ' Entisols'.

Peaty and Marshy Soils

- These soils occur in humid regions with accumulation of high organic matter. During monsoons the soils get submerged in water and the water receipts after the monsoon during which period rice is cultivated. Soils are black clay and highly acidic with pH of 3.5. Free alluminium and ferrous sulphate are present.
- The depressions formed by dried rivers and lakes in alluvial and coastal areas some times give rise to water logged soils and such soils are blue in colour due to the presence of ferrous iron.
- Peaty soils are found more in Kerala and marshy soils are found more in coastal tracks of Orissa, West Bengal and South - East coast of Tamilnadu.

Saline - Sodic Soils

- Saline soils contain excess of natural soluble salts dominated by chlorides and sulphates which affects plant growth. Sodic or alkali soils contain high exchangeable sodium salts.
- Both kinds of salt effected soils occur in different parts of India like Uttarpradesh, Haryana, Punjab, Maharastra, Tamilnadu, Gujarat, Rajastan and Andhra pradesh. These soils are classified under ' Aridisols', ' Entisols' and ' Vertisols'.

SOIL CONSTITUENTS AND PROPERTIES

In terms of its constituents, the soil can be viewed as a three-phase system comprising solid, liquid and gaseous constituents. The solid phase consists of both mineral and organic material; the mineral fraction is derived largely from the parent material and the organic fraction largely from vegetation growing in and above the soil. Individual particles and fragments often join together to form larger units known as *aggregates* or *peds,* and these play a crucial role in the development of many physical and chemical characteristics of soils. Between the solid material there are usually spaces known as *pores* or *voids,* and these are occupied by the liquid and gaseous phases. The liquid component, or *soil water,* derived from precipitation and ground-water sources, is able to transport material through the soil in both suspended and dissolved forms, and is often referred to as the *soil solution.* The gaseous component, known as the *soil atmosphere* or *soil air,* consists of a mixture of gases derived from the above-ground atmosphere and from the respiration of soil organisms. The relative proportion of constituents in a typical topsoil is shown in Figure, although the proportions can vary widely depending on the time of sample collection, soil type and environmental conditions.

The constituents are intimately associated and interact in a variety of ways, which gives a soil a wide range of properties, from relatively simple morphological characteristics recognisable in the field to complex chemical properties recognised only by laboratory analysis, and many of the properties are strongly interrelated. Although technical and analytical procedures will be examined briefly, more specialised texts should be consulted for further details.

SOIL CONSTITUENTS

Mineral Material

The mineral fraction of soils is derived largely from weathering of the underlying parent material, which may consist of consolidated bedrock or unconsolidated superficial deposits. Consolidated bedrock can be classified into one of three categories—igneous, sedimentary or metamorphic. Igneous rocks are derived from the consolidation of molten magma, either within the Earth's crust or at its surface, whereas sedimentary rocks are the product of cycles of weathering, erosion and deposition operating at the surface, and metamorphic rocks originate from the alteration of any other rock type by high temperature and/or pressure but without melting. Unconsolidated superficial deposits are just as variable as solid bedrock materials and are often classified according to the nature of their depositional environment, for example riverine alluvium, glacial till and lacustrine clays. These deposits are often indurated or cemented to some extent by calcareous, siliceous or iron-rich components. In addition to the parent material origin of mineral material, it can also be added to soils by movement from upslope, by aeolian transport or by atmospheric fall-out of materials such as volcanic ash.

Soil minerals occur in a wide variety of forms, but these can be arranged into groups on the basis of their chemical composition and the structural arrangement of their constituent elements. By far the most abundant group in soils and their parent materials is the *silicate* group. The fundamental building block of all silicate minerals is the silicon-oxygen (Si-O) tetrahedron, which consists of a central silicon ion (Si^{4+}) surrounded by four closely spaced oxygen ions (O^{2-}). Because the silicon ion has four units of positive charge and each oxygen ion has two units of negative charge, each discrete tetrahedron possesses four units of negative charge. This allows the tetrahedra to link together in a variety of characteristic structural arrangements which forms the basis of their classification. The potential range of silicate minerals is further widened by substitution of one ion for another within the mineral structure, a process known as *isomorphous substitution*. One of the most common substitutions involves replacement of some Si^{4+}, in Si-O tetrahedra, by aluminium ions (Al^{3+}). Other examples include the substitution of Al^{3+} by magnesium ions (Mg^{2+}) or iron (Fe^{3+} or Fe^{2+}). When ions with the same charge or valency are exchanged, the mineral structure remains electrically neutral. If, however, ions with different valencies are exchanged, there will be a charge imbalance; frequently this results in an excess negative charge, as in the examples above. In this situation, electrical neutrality is achieved either by incorporation of additional cations such as calcium, magnesium, potassium or sodium (Ca^{2+}, Mg^{2+}, K^+, Na^+) into the crystal lattice, or by structural rearrangements that allow internal compensation of charge. The most common silicate minerals and their associated structures are discussed briefly below.

Framework silicates or tectosilicates consist of a three-dimensional lattice of Si-O tetrahedra linked through their corners. The two main mineral groups within this category are quartz and feldspars, which are common minerals in many soils. Quartz consists simply of Si-O tetrahedra linked through the O^{2-} ions, and consequently there are twice as many O^{2-} ions as Si^{4+} ions in the mineral structure, which has the general formula $(SiO^2)n$. Unlike quartz, feldspars contain significant amounts of Al^{3+} ions, originating from isomorphous substitution of Si^{4+} ions, and base cations, especially Ca^{2+}, Na^+ and K^+, which satisfy the residual negative charges. There are two main groups of feldspar minerals—potassium feldspars such as orthoclase and microcline, and plagioclase feldspars which form a continuous series of minerals between the sodium end member, albite, and the calcium end member, anorthite.

Chain silicates or *inosilicates* comprise Si-O tetrahedra linked together to form a continuous chain structure of which there are two types. First there is the single chain where the Si-O tetrahedra are linked by sharing two out of the three basal O^{2-} ions, and second there is the double chain where tetrahedra are linked by sharing all three basal O^{2-} ions to form a hexagonal arrangement. Common minerals in these two categories are the pyroxene family and the amphibole family respectively. In chain silicates the chains themselves are linked together by a variety of cations including Ca^{2+}, Mg^{2+}, Fe^{2+}, Na^+ and Al^{3+}. Minerals in this group are very reactive and more easily weathered than the framework silicates. *Orthosilicates* and *ring-silicates* display a wide range of structures but, in comparison with other groups, their occurrence in soils is relatively minor. Some minerals in this group are very reactive and therefore particularly susceptible to weathering, whereas others are very resistant. This category can be divided into two groups—*neso-silicates* and *soro-silicates*. In the first group the Si-O tetrahedra occur as separate units, with no shared O^{2-} ions, and are linked by metallic cations. Examples include olivine and garnet. In soro-silicates the tetrahedra form separate groups in which they share one or more of their O^{2-} ions; when the group is formed by sharing O^{2-} ions by more than two Si-O tetrahedra, a ring structure results. Examples of soro-silicates are beryl and cordierite.

Sheet silicates or *phyllosilicates* are perhaps the most important group of minerals in soils, playing an important role in the development of many physical and chemical soil characteristics. Unlike the other mineral groups, many sheet silicate minerals occur in the smallest grain-size categories; consequently they are often known as the *clay minerals*. Minerals in this group are made up of various combinations of three fundamental sheet structures: (a) Si-O (siloxane) sheet in which the Si-O tetrahedra are linked in a hexagonal arrangement, (b) Al-OH (gibbsite) sheet in which the Al^{3+} ions are surrounded by six closely packed hydroxyl (OH^-) ions to form an octahedral structure, and (c) Mg-OH (brucite) sheet whose structure is similar to that of the gibbsite sheet but contains Mg^{2+} ions rather than Al^{3+}.

Classification of sheet silicate minerals is based on three criteria—the ratio of the above sheets within a unit layer of the mineral structure, the interlayer or basal spacing between unit layers, and the interlayer components or species. The sheet silicate minerals most commonly found in soils include the micas, illite, chlorite, kaolinite, vermiculite and the smectites. The most frequently occurring micaceous minerals are muscovite and biotite. Both of these are 2:1 layer silicates with an interlayer spacing of approximately 10 Å (1.0 nm) and K^+ as the dominant interlayer component. Illite is very similar in composition and structure to micaceous minerals and is sometimes known as hydrous mica.

Chlorite has a 2:1:1 layer structure (two siloxane sheets to one brucite sheet, forming a biotite mica layer, to one further brucite sheet which occupies the interlayer position) with an interlayer spacing of about 14 Å (1.4 nm). Kaolinite has a 1:1 layer structure and an interlayer spacing of around 7 Å (0.7 nm). Layer units of kaolinite are attracted by hydrogen bonding and the crystals have a characteristic hexagonal appearance. Vermiculite is a 2:1 layer silicate with a basal spacing of about 14 Å (1.4nm) and interlayer positions occupied by Mg^{2+} ions and water molecules. The most common smectitic mineral in soils is montmorillonite. This is a 2:1 layer silicate with interlayer positions dominated by Na^{+} and Ca^{2+} ions, and water molecules. Although its basal spacing is approximately 14 Å (1.4 nm), this varies somewhat due to its shrink-swell characteristics which occur on wetting and drying.

Other sheet silicate minerals sometimes found in soils include halloysite, imogolite and allophane, and mixed layer or interstratified minerals. Halloysite is similar in many ways to kaolinite, but its crystal layers are curved to form a tubular structure. Imogolite and allophane are gel-like hydrated alumino-silicate minerals with a poorly crystalline tube- or thread-like structure. Mixed layer minerals result from the interlayering of more than one sheet silicate mineral. Interlayering may be regular, where there is a regular repetition of the different mineral layers, or random where there is no clear pattern of repetition. These minerals may exist as intergrades which represent stages of transition between one mineral and another during weathering. Examples of mixed layer minerals are mica- or illite-vermiculite (hydrobiotite), illite-montmorillonite and kaolinite-montmorillonite.

In addition to the silicate minerals are the *non-silicate minerals*. These are often of only minor occurrence in soils and are usually referred to as *accessory minerals*, although they play a significant role in the development of some soil characteristics. Among the most common non-silicate minerals in soils are the free oxides and hydroxides of iron, aluminium and manganese; these may exist as crystalline or amorphous forms. Iron minerals include goethite, ferrihydrite, hematite, lepidocrocite, maghemite and magnetite. Aluminium minerals include gibbsite and boehmite, while examples of manganese minerals are birnessite and pyrolusite. Other non-silicate minerals include calcite ($CaCO_3$), which occurs in soils developed from calcareous parent materials, and anatase (TiO_2) and amorphous silica, which are often found in soils developed on recent volcanic deposits.

Organic Components

Soil organic matter is derived from a number of sources, of which the most important is usually plant litter. This consists of a variety of plant debris including leaves, stems, flowers, twigs, bark, and the larger branches and trunks of trees. Other organic components include plant roots, root exudates, soil organisms together with their faecal remains and metabolites, and organic substances washed into the soil from vegetation.

Soil organisms are responsible for the physical comminution and bio-chemical decomposition, and incorporation of organic matter. They exhibit tremendous diversity in terms of their numbers, size and morphology, and also vary dramatically in their function, mode of nutrition and environmental tolerance. Organisms can be referred to as *producers, consumers* or *decomposers*. Producers, such as plants, fix carbon from atmospheric carbon dioxide (CO_2) during photosynthesis, consumers feed on plants and other organisms, and decomposers utilise carbon from organic material, returning it to the atmosphere as CO_2 and other gaseous by-products and mineralising nutrients to their original ionic form. Soil organisms can also be classified according to their size *micro-organisms,* which include both microflora and microfauna, *mesofauna* and *macrofauna*. The classification of soil organisms according to their mode of nutrition is based on their sources of carbon and energy. Essentially, they obtain carbon from either organic material or inorganic sources (largely CO_2); these two groups are known as *heterotrophs* and *autotrophs* respectively. They can also obtain energy either from light (photoheterotrophs or photoautotrophs) or from chemical oxidation (chemoheterotrophs or chemoautotrophs). In addition to their carbon and energy sources, soil organisms, especially bacteria, vary in their oxygen requirements. *Aerobes* have a direct requirement for oxygen, *facultative anaerobes* normally require oxygen but may adapt to oxygen deficit by utilising nitrate and other inorganic compounds as electron receptors, and *obligate anaerobes* require an absence of oxygen as it is toxic to them. Before looking at the composition of organic material, the various groups of micro-organisms, mesofauna and macrofauna will be briefly considered.

Four main groups of micro-organisms can be recognised—bacteria and actinomycetes, fungi, algae and protozoa. Bacteria are very small organisms and are often round or rod-like in shape. They tend to exist in thin films of water surrounding soil particles, often reproducing with tremendous rapidity, with numbers in the range of $10^6 - 10^9$g having been widely reported. Many species secrete polysaccharide gums, which facilitate aggregation. Bacteria demonstrate great biochemical versatility in their ability to decompose a wide range of materials under a variety of conditions. For example, *Pseudomonas* sp. can metabolise a range of chemicals including pesticides, *Nitrobacter* sp. derives its energy from the oxidation of nitrite to nitrate, *Thiobacillus ferrooxidans* acquires energy from oxidation of reduced sulphur compounds and Fe^{2+}, and *Rhizobium* sp. forms nitrogen-fixing nodules on the roots of leguminous plants. Actinomycetes consist of fine branching filaments (about 1 ìm in diameter) during their vegetative stage, when they are similar in appearance to fungi. During reproduction, however, the filaments undergo fragmentation, sometimes forming dense colonies. For this reason actinomycetes are often classified as highly evolved and complex bacteria.

Fungi produce filamentous structures (hyphae) which are about 0.5-1.0 ìm in diameter and which grow into a dense network or mycelium. They are generally less numerous than bacteria in soils ($1 - 4 \times 10^5$ g),.although they are common in acidic soils where they can be responsible for 60-80 per cent of organic matter decomposition. All fungi are heterotrophic, living mostly in the surface layers where they play an important role in the development of soil structure. Mycorrhizal fungi live symbiotically in plant tissue, removing carbon in return for nutrients such as phosphorus. Algae are photosynthetic organisms and are therefore confined largely to the soil surface. They include Cyanophaceae (blue-green algae) and Chlorophaceae (green algae), the former playing a major role in development of the nitrogen status of soils. Algal numbers are usually intermediate to those of bacteria and fungi at around $10^5 - 3 \times 10^6$ g. Protozoa are microfauna, unlike the previous groups which are microflora. They are uni- or non-cellular organisms of 5-40 ìm in length and live in thin water films surrounding soil particles. They can be divided into amoeboid forms with silica or chitin sheaths, and rotifers with better differentiated cell structure. Protozoa are very numerous (often > 10^4 g) and are particularly important in controlling the numbers of bacteria and fungi, on which they feed.

As in the case of micro-organisms, four main groups of mesofauna can be recognised—nematodes, arthropods, annelids and molluscs. Nematodes are unsegmented worms, and next to protozoa are the smallest of the soil fauna, being 0.5-1.0 mm in length. They are plentiful in soil and litter, feeding on plant remains, roots, bacteria and sometimes on protozoa. Arthropods can be divided into a number of categories, the most significant numerically being acari (mites) and collembola (springtails). Both of these feed on plant litter, bacteria and fungi, with acari being particularly common in acidic litter where they may constitute 80 per cent of soil fauna. Other arthropod groups include myriapods, which are dominated by chilopods (centipedes) which are carnivorous, and diplopods (millipedes) which are herbivorous. Arthropods also include isopods (woodlice), beetles, insect larvae, ants and termites. Termites are prevalent in tropical soils and are particularly active soil mixers, producing structures at the surface (termitaria) up to several metres in height.

Arthropod populations can be particularly high, for example 220,000 m^2 in old grassland soils.

Annelids consist largely of enchytraeid worms (potworms) and lumbriscid worms (earthworms). Enchytraeids are small (0.1 – 5.0 cm in length) and have a thread-like appearance. They feed on algae, fungi, bacteria and soil organic matter, and can occur in populations as high as 200,000 m^2. Lumbriscid worms are probably more important than any other soil invertebrate in the decomposition and mixing of organic matter, and their numbers can exceed 800 m^2, although they do not tolerate highly acidic conditions, and during prolonged drought or heat they burrow deeply into the soil and become inactive. Some species, such as *Lumbricus rubellus,* live only in the surface litter layer, but most migrate between organic and mineral horizons, ingesting both constituents and ensuring thorough mixing of the soil. Large populations under grassland can consume 90 ha, and the casts which are produced are important in the development of soil structure. The main mesofaunal mollusc groups are slugs and snails (gastropods) which feed on plants, fungi and faecal remains, but these are often limited in biomass to 20-45 g/m^2. Macrofauna include larger molluscs, beetles and larger insect larvae, together with larger vertebrates such as moles, rabbits, foxes and badgers. These often burrow deeply into the soil, feeding on smaller organisms in their path. Moles in particular are voracious feeders and consume large numbers of earthworms, insect larvae and slugs.

In terms of its composition, organic material consists predominantly of carbon, hydrogen, oxygen and nitrogen, with carbon providing the framework for organic structures; most organic residues in soils contain around 45-55 per cent carbon by weight. The elements which constitute organic material are arranged to form a variety of compounds, some of which have very complex structures. The basic building block of many organic compounds, however, is the carbon tetrahedron, where a central carbon atom with a valency of four links with other elements, notably hydrogen, oxygen and nitrogen; this structural arrangement is very similar to the basic building block of silicate minerals. In some instances, numerous tetrahedra link together to form extremely large molecules known as *polymers*. Linkages can produce both linear and cyclic molecules, and the number of potential combinations and structural arrangements is almost endless. Consequently, organic structures and their variety are considerably more complex than those of mineral material and are less well understood, although the majority of organic compounds can, however, be isolated. The main groups of compounds are carbohydrates, proteins and amino acids, and lignin, along with smaller amounts of fats, waxes, pigments and resins. The proportions of these compounds vary according to the type of organic matter and its stage of decomposition.

Essentially, carbohydrates are hydrates of carbon with a general formula $C \times (H_2O)y$, although some contain other elements such as nitrogen and sulphur. Carbohydrates are the main constituent of plant material at 60-90 per cent of dry mass, and in soils 5-30 per cent of the carbon exists in this form. The simplest and most easily broken down carbohydrate compounds in soils are the sugars and starches. These comprise relatively small molecules such as glucose (a simple sugar), which is a monosaccharide. More complex, and thus more resistant to breakdown, are hemi-cellulose (a pentose sugar polymer) and cellulose (a β (1-4) glucose polymer). These are polysaccharides which consist of monosaccharide units joined by C–O–C links. Cellulose is the most abundant carbohydrate in plants, where it may constitute more than 40 per cent of the carbon. It forms the resistant fibres of plants, the breakdown of which is catalysed by enzymes (cellulases) secreted by micro-organisms. Breakdown ultimately converts the polysaccharide compounds into monosaccharide compounds which can then be assimilated directly by the micro-organisms. Chitin is a polysaccharide of similar composition to cellulose, but it also possesses some nitrogen in the form of amino groups. It is a common constituent of insect cuticles and is also often present in fungi.

Proteins and amino acids are composed of about 50-55 per cent carbon, 20-25 per cent oxygen, 15-20 per cent nitrogen and 6.5-7.5 per cent hydrogen, although some contain small amounts of phosphorus and sulphur. They are an important source of nitrogen, with around 20-50 per cent of all organic nitrogen in soils existing as amino acids. Although they are rapidly metabolised in soils, they can persist for long time periods due to adsorption onto other constituents, such as clay, or by combination with more resistant organic components such as lignin or tannin. Lignin is a particularly resistant component of soil organic material, and constitutes 15-35 per cent of the supporting tissues of plants. Essentially it is a complex phenolic polymer which displays a high degree of aromaticity in the development of cyclic, benzene ring structures. The aromatisation and large number of C-C bonds are the main cause of its resistance to decomposition.

As a result of organic matter decomposition, fresh organic components are gradually converted, via a range of fermented products, to a material known as *humus,* which represents the end-product of this process. Humus is finely divided and amorphous with no cellular structure. Essentially it consists of insoluble, heterogeneous polymers and has an approximate composition, on an ash-free basis, of 44-53 per cent carbon, 40-47 per cent oxygen, 3.5-5.5 per cent hydrogen and 1.5-3.5 per cent nitrogen.

Water

Soil water is derived from two principal sources—precipitation and ground water. Precipitation arrives at the soil surface in various forms; although rain, snow and hail are the main types, fog and mist may provide significant amounts of moisture, particularly in coastal and upland areas. The proportion of precipitation that reaches the ground surface depends largely on the nature and density of vegetation cover. On surfaces devoid of vegetation, precipitation reaches the soil directly, but on vegetated surfaces a significant proportion of precipitation can be intercepted; much of this ultimately reaches the soil as canopy throughfall and stemflow, while some is returned to the atmosphere by evapouration. On reaching the surface, water can either infiltrate the soil or, if the rate of arrival of water at the surface exceeds the rate of infiltration, it will run off over the surface. The precipitation that infiltrates the soil will either be lost via evapotranspiration or drainage, or will be retained by forces which hold it in place or allow it to move only very slowly. Ground water can be derived by lateral movement from upslope, or by upward movement from the underlying rock strata.

The composition of soil water is a particularly dynamic characteristic, varying dramatically even over short time periods. This behaviour arises from the intimate association between the water, small mineral and organic particles (clay and humus) and plant roots, which can involve the exchange of ions between these components. Soil water contains a number of dissolved solid and gaseous constituents, many of which exist in mobile ionic form, and a variety of suspended solid components. Base cations (Ca^{2+}, Mg^{2+}, K^+, Na^+, NH^{4+}) may be derived from a number of sources. They are present in the atmosphere and are later dissolved in precipitation; in maritime areas, for example, precipitation often contains significant quantities of marine salts which are particularly rich in sodium and chloride ions (Na^+ and Cl^-). As well as being deposited in the soil directly, ions accumulate on the surfaces of vegetation and are subsequently washed off into the soil. Cations can also be derived from mineral weathering and organic matter decomposition, entering the soil directly or via the soil exchange system; these processes play an important role in the buffering of soil acidity. In agricultural systems, lime and fertilisers provide yet another source of base cations, particularly Ca^{2+}, K^+ and NH^{4+}.

The concentration of hydrogen ions (H^+) in soil water is a measure of its acidity, which is expressed in terms of *pH*. A major source of such acidity is carbon dioxide (CO_2) which is derived from the atmosphere, where it is dissolved in precipitation, and from the soil air where it is a product of soil organism respiration. Carbon dioxide dissolved in water Unpolluted rain water in equilibrium with atmospheric CO^2 has a pH of about 5.6, whereas soil water in equilibrium with CO_2 in soil air is usually more acidic, with pH values often below 5.0; this is because CO_2 levels in soil air are considerably greater than in the atmosphere. Another source of acidity in soil water derives from industrial and urban emissions. In addition to these pollutants, organic acids derived from decaying organic material are an important source of soil acidity. H^+ is also released by plants in exchange for nutrient base cations, and as part of the process of nitrification where NH^{4+} is converted to NO^{3-}.

Iron and aluminium are also important constituents of soil water, particularly under acidic conditions. They are derived from mineral weathering and may exist in the soil solution either as ions (Fe^{2+} and Al^{3+}) or in the form of soluble organo-metallic complexes; significant quantities of dissolved organic carbon may also exist in this form. In some circumstances aluminium may be mobilised by mineral acids, such as sulphuric and nitric acid deposited in acid precipitation.

As in the case of cations, the anions in soil water are derived from a number of sources. Nitrate and phosphate ions (NO_3 and PO_4^{3-}) are produced by mineralisation processes during organic matter decomposition, and are also derived from fertilisers. Chloride ions (Cl^-) and, to a lesser extent, sulphate ions ($SO_4{}^{2-}$) originate from atmospheric sources, including airborne marine salts and acid deposition. Bicarbonate ions (HCO_3) originate largely from the dissociation of HCO_3 as mentioned above, and are associated with mineral weathering, especially in soils developed from carbonate-rich parent materials.

In addition to the major dissolved constituents of soil water, there are other dissolved components although these are usually relatively minor and local in their occurrence. These include organic material and silica, together with a number of pollutants such as heavy metals and radionuclides.

Soil water contains not only dissolved solids but also a number of suspended constituents. These include small particles of mineral and organic material, which often result in discolouration and increased turbidity of soil water. Similarly, precipitates may accumulate in soil water, usually as a result of chemical changes as the water migrates through the soil. For example, orange-brown precipitates of insoluble ferric iron compounds can sometimes be observed where water, in which iron compounds usually exist in the soluble ferrous form, becomes more oxygenated. The chemical characteristics of water can also influence the behaviour of fine sediments, particularly clays, in suspension. In waters with low solute concentrations, or significant quantities of monovalent cations such as Na^+, the clays tend to remain in a dispersed state and can be held in suspension for long periods of time. In waters with high solute concentrations, however, particularly where concentrations of divalent cations such as Ca^{2+} are high, the clays may undergo rapid flocculation and settling.

Air

Air and water have a reciprocal arrangement in terms of their occupancy of soil pore space; in saturated soils, air content is low, whereas in dry soils the pore spaces are largely air-filled. Changes in water and air content are particularly dynamic because much of the water present in a saturated soil drains away rapidly, while heavy rainfall can quickly bring the soil back to saturation. The gaseous constituents of soil air are derived largely from the atmosphere, the respiration and metabolism of soil organisms, and from the evapouration of soil moisture. Soil air is continuous with the atmosphere provided that the soil surface is not sealed due to compaction or crusting, and such continuity ensures the free movement and exchange of gases. The gases move along gradients of partial pressure, so that oxygen will tend to migrate from the atmosphere where its partial pressure is high into the soil where it is low. Conversely, carbon dioxide and water vapour will tend to migrate from the soil into the atmosphere. Movement of gases may occur by diffusion, mass flow or in dissolved form, with diffusion being by far the most significant of these processes.

The atmosphere contains approximately 78 per cent nitrogen, 21 per cent oxygen and 0.03 per cent carbon dioxide by volume. In comparison, soil air contains similar amounts of nitrogen, slightly less oxygen and more carbon dioxide. The balance between levels of oxygen and carbon dioxide in soil air depends largely on the rate of respiration of soil organisms and on the diffusivity characteristics of the soil. Respiration rates often vary seasonally in response to the availability of organic substances for consumption, and to variations in temperature and moisture conditions. Diffusion of gases occurs most effectively in dry soils with large and interconnected pores. The presence of water tends to reduce pore continuity and consequently the soil becomes increasingly oxygen deficient or *anaerobic*. In addition to carbon dioxide, organisms release other gases into the soil, including methane (CH_4) and hydrogen (H_2), as a result of organic matter decomposition.

MINERAL PARTICLES

The principal properties of soil mineral particles in an environmental context are their size, shape, nature of surface, orientation and mineralogy. The mineral fraction of soils consists of particles that vary dramatically in size from large boulders, through cobbles and pebbles to sand, silt and clay. Most studies of soil are concerned with the <2 mm size range, which is often referred to as the *fine fraction* or *fine earth*. It is the proportion by weight of the size categories within the fine fraction which defines the particle size distribution or *texture* of a soil. The particle size distribution of a soil is usually recorded numerically on a per centage by weight basis, or graphically using either a triangular graph, on which only three size categories can be shown, or a cumulative frequency curve which allows a fuller range of size classes to be recorded. Most soils comprise a continuous spectrum of particle sizes, and the width of this spectrum is defined by the degree of *sorting*. Poorly sorted soils possess a wide range of particle sizes, whereas well sorted soils have a narrow range. Texture can be estimated in the field simply by rubbing the soil between thumb and forefinger. Sand grains are easily distinguished by their coarseness, while silt has a distinctive soapy feel and clay is characteristically plastic and mouldable when moist.

The morphology of mineral particles relates to their shape and surface characteristics. For large particles this can be determined in the field, but for smaller material it is necessary to use a microscope. Particle shape can be expressed in two or three dimensions. Common terms adopted in two-dimensional expression include roundness, angularity and elongation, while in three-dimensional expression particles are recorded in terms of the extent of sphericity and flatness, using descriptions such as sphere, disc, rod, cube and prism. Such descriptions can be qualitative, or by reference to a comparison chart semi-quantitative estimates can be made. A number of indices have also been devised in order to quantify shape more accurately. The nature of the surface of a particle is usually referred to as its *surface texture,* although this in no way relates to the texture of a soil as an expression of its particle size distribution. Surface texture is most commonly evaluated for the fine fraction, using a scanning electron microscope, which allows the grain surfaces to be viewed in three dimensions at very high magnification.

Orientation of mineral particles describes the disposition of the particle in three dimensions, and this can be a useful indicator of the direction of movement of soil material. The measurement of orientation is usually made with reference to particle longest axes, either in the horizontal plane as a compass bearing, or in the vertical plane as an angle of dip or plunge. In some cases, both sets of measurement are made and the results combined. Orientation can be depicted in a variety of ways including bar charts, rose diagrams and polar scattergrams.

Minerals differ markedly in their composition and consequently many physical and chemical characteristics of soils, including texture, acidity and nutrient status, can be related to their mineralogy. Mineralogical studies are also important in the examination of weathering in soils. The mineralogical characteristics of large particles can be evaluated in the field by the inspection of hand specimens, but for smaller particles laboratory methods are required. Sand- and coarse silt-sized particles are usually examined using a petrological microscope, but clay and fine silt are too small to be easily resolved in this way and their mineralogy is commonly identified using the technique of X-ray diffractometry. This allows the crystal lattice dimensions to be measured and is particularly useful in the identification of sheet silicate minerals. Samples are scanned by an X-ray beam and when dominant lattice spacing is encountered, diffraction occurs and a peak is registered on a moving chart recorder. Each mineral has a characteristic set of peaks, although in some cases more than one mineral may have a peak in the same position. However, it is possible to distinguish between these minerals using simple chemical or heat pre-treatments which produce characteristic changes in the lattice spacing of one mineral but not in another.

Aggregates

Aggregation in soils is promoted by a number of physical, chemical and biotic forces. Physical forces include expansion and shrinkage associated with wetting and drying, and compaction by raindrop impact, animal trampling and agricultural machinery. Chemical forces are largely electrostatic in character and often depend on the presence of adsorbed cations in association with the negative surface charge of colloidal particles such as clay and humus. The generation of negative surface charge on colloidal particles and the processes of cation adsorption are discussed. Multivalent cations in particular, such as Ca^{2+}, Mg^{2+} and Al^{3+}, have the ability to form an attachment with more than one colloidal particle, a process known as *cation bridging*. Similarly, attraction may occur between positive charges on the broken edges of sheet silicate particles and the negative charges on the faces of other similar particles. In soils with significant positive charge, anion bridging may occur, particularly if multivalent anions are present. In response to the various mechanisms of interparticle attraction, particles flocculate together to form larger units known as *domains;* these may be up to 5 ìm in diameter. In addition to electrostatic interparticle forces, interaggregate attraction can occur due to the binding effects of various organic compounds, fungal hyphae and plant roots. Organic polymers, such as polysaccharide gums, together with organic mucilages and fungal hyphae, facilitate the binding of domains to form microaggregates which may be up to 250 ìm in diameter. At the larger scale, plant roots and fungal hyphae play an important role in the binding of microaggregates to form macroaggregates or *peds* which can be easily observed in the field. Inorganic cementing agents such as carbonate and iron compounds can perform a similar binding function in soils.

Because of the different scales of aggregation, it is often viewed in terms of a hierarchical model comprising building blocks which increase progressively in size. The degree of persistence of aggregates varies between the different levels in the model. Electrostatic forces predominate at the domain level and are particularly resistant to change. Similarly, at the microaggregate level, binding forces are relatively persistent, although they vary to some extent according to the organic content of the soil. In contrast, at the macroaggregate level, binding forces are often transient, being closely influenced by variations in plant cover and the development of root networks.

Aggregates, or peds, which persist during wetting/drying and freezing/ thawing cycles form the basis of soil *structure*. Soil structure is defined in terms of the size, shape and arrangement of particles, aggregates and pores. It is classified on the basis of ped morphology, class and grade. There are four main groups of ped morphology—spheroidal, blocky, prismatic and platy. Spheroidal peds are equidimensional in form and can be divided into granular and crumb types. Granular and crumb structures are most commonly found in the A horizons of soils, their development being facilitated by the presence of roots and other forms of organic material. Blocky peds are also equidimensional and may possess several curved or planar surfaces. Frequently they are observed in B horizons, particularly in soils with significant amounts of clay, where they develop in response to the effects of regular wetting and drying cycles. Prismatic and associated columnar peds are characterised by strong vertical and weak horizontal development. The tops of prismatic peds are poorly defined, whereas those of columnar peds are often rounded in form. Like blocky structures, prismatic and columnar structures often develop as a result of wetting and drying but are found in the B horizons of less clay-rich soils. Platy structure displays strong horizontal and weak vertical development, and often develops as a result of compaction. The structures are sometimes lenticular in form, with their centres being thicker than their edges, and these are often found in soils that are subjected to regular freezing and thawing cycles. Peds which comprise more than one structural type are often observed in soils and are known as *compound peds*. For example, large prismatic peds may be made up of smaller, blocky peds. It is also not uncommon to find that structures vary between the horizons within a soil profile, such that the A, B and C horizons could possess crumb, prismatic and blocky structures respectively.

Ped class or size is usually classified according to the system shown in Table. Grade, which describes the distinctiveness and durability of peds, may be expressed as structureless, or as weakly, moderately, well or strongly developed. An apedal soil is described as massive if it is coherent and as single grain if it is not. Usually, massive soils are fine-textured while single grain are coarse-textured. Weakly developed structure comprises poorly formed, indistinct and weakly coherent peds, most of which will be broken down if the soil is disturbed, while strongly developed structure comprises well formed, distinct and durable peds. The grade of structure depends on a number of soil characteristics, particularly moisture content and the quantity of aggregating colloids such as clay and humus; peds tend to be considerably more durable in dry soils and in those with high colloidal contents.

Pore Space

Pore spaces vary dramatically in shape from spherical voids to tortuous, interconnecting cracks and channels. They also vary in size from large macropores of several cm in diameter to very fine micropores which may be <1 ìm in diameter. Generally, a diameter of 60 ìm (0.06 mm) is taken as the size boundary between macro- and micropores. Pore space will influence both the *bulk density* and the *porosity* of a soil. Bulk density refers to the specific gravity of a bulk soil sample, usually collected as an undisturbed core. Its calculation is then derived from measurement of the mass and volume of the dried core. Values of bulk density are generally considerably lower than those of the particles which make up a soil, because they are based on both solid material and pore space rather than on the solid components alone. For example, the average density of soil particles is often assumed to be 2.65 g/cm^3, whereas bulk density values can range from around 2.0 g/cm^3 in sandy soils and compacted clay layers to <1.0 g/cm^3 in organic soils. If the bulk density of a soil is not measured, then an average value of 1.33 g/cm^3 is often assumed. In addition to porosity, pore size distribution is an important soil characteristic. It is associated directly with water retention, drainage and aeration, and therefore has a major influence on plant growth. It is most often established from moisture characteristic or moisture retention curves. This involves determining the volumetric water content at various points over a range of tensions or suctions applied to an undisturbed soil core. As tension is increased, water is removed from progressively smaller pores and because tension is inversely proportional to pore radius, the volume of pores of a certain size can be determined from the amount of water extracted at the appropriate tension. In Figure, for example, the total porosity is 50 per cent and, of this, 15 per cent of pores are > 15 ìm in diameter, 15 per cent are 1.5-15 ìm, 10 per cent are 0.15-1.5 ìm and 10 per cent are <0.15 ìm.

Porosity and pore size distribution are influenced by a number of soil characteristics, particularly texture, degree of aggregation, bulk density, presence of swelling clays, and organic content. Closely packed sands, for example, could have a theoretical minimum porosity of about 10 per cent. However, as soils usually possess a wide range of particle sizes and aggregates, and because soil particles are rarely spherical, values of porosity are usually much greater than this. Sandy and compacted clay soils may have porosities of <40 per cent, whereas a fine-textured A horizon can have values of > 60 per cent.

Moisture

Soil water possesses free energy which is a measure of its potential for movement and change in the soil. In soils with a high moisture content, forces attracting the water to solid particles are weak and its free energy is high. As moisture content decreases, however, the attractive forces become progressively stronger and its free energy decreases. Soil moisture is affected by three types of force determined by soil properties which can either encourage or restrict water movement. First is *adsorption* whereby water molecules are attracted to the surfaces of colloids mainly by electrostatic forces. This is therefore an important process in soils with high clay or organic matter contents, in which such forces are high. Second is *capillarity* by which water is held in soil pores by adsorptive forces between the water and pore surfaces and surface tension forces at the water surface. The force by which the water is held increases with decreasing pore size, therefore water can drain out of large pores more easily than from smaller ones. The combined effect of adsorptive forces and capillarity is known as *matric suction*. The third force is *osmosis* which occurs between solutions of different ionic concentrations, water moving from lower to higher concentration solutions. This is important in saline soils in which high solute concentrations can form due to the relative ease of dissolution of salts.

These forces are usually expressed in units of atmospheres (atm), bars or kiloPascals (kPa), with 1 atm being approximately equal to 1 bar, and 1 bar equal to 100 kPa. In the past, these forces were sometimes expressed as the logarithm of the *hydraulic head* (pF); this is the length of a column of water which is required to produce a given positive pressure, or which can be supported by a given negative pressure (tension). This concept is, however, rarely used today.

Soil water is held most strongly when it is adsorbed onto colloidal particle surfaces in the form of thin films only a few molecules in thickness, or when it is bound up in mineral structures (structural water). Further away from particle surfaces, water is held in small micropores at tensions of 0.05-30 atm by capillarity; much of this water can be lost from the soil through evapouration and plant uptake. Water held at tensions below 0.05 atm occurs in the macropores, and this will usually drain rapidly from the soil, in two or three days, under the influence of gravity and is therefore known as *gravitational water*. Thus, on moving away from particle surfaces into progressively larger pores, tensional forces become progressively weaker, decreasing logarithmically with increasing pore size, until they are unable to counteract the effect of gravity. A soil which has lost all of its gravitational water through drainage contains the maximum level of plant-available water and is said to be at *field capacity*. If, however, a soil loses all of its available water through evapouration and plant uptake, without further wetting, then it is said to have reached *wilting point*.

A variety of methods are available for the measurement of soil moisture. For example, it can be determined gravimetrically using bulk samples; these are weighed in their field-moist and oven-dried states, the weight difference representing the moisture content, which is then expressed as a percentage of either field-moist or oven-dried soil. Alternatively, moisture content can be expressed on a volumetric basis using soil cores of known volume. Weight measurements are made as above and, given that the specific gravity of water is 1.0 g/cm^3, the moisture content can then be expressed as a percentage of the soil volume. Changes in moisture content within a soil can be determined using a neutron probe. As the probe is lowered through aluminium tubes inserted into the soil, it emits neutrons which collide with the hydrogen nuclei contained in water molecules, and a sensor detects backscatter from the collisions, the intensity of which is in direct proportion to water content. Problems have, however, been experienced with the use of neutron probes, particularly in soils containing significant amounts of swelling clays, where cracking results from drying and associated shrinkage.

The strength of tensional forces (matric potential) by which water is held in the soil can be determined using a tensiometer. This consists of a sealed plastic tube filled with water, with a ceramic cup at one end and a vacuum gauge at the other. The tube is set in the soil at the required depth and the gauge adjusted to zero. As water moves out into the soil through the cup, tension increases progressively until the water in the tensiometer is in equilibrium with that in the soil. At this point, movement ceases and the tension recorded on the gauge is a measure of the matric potential; tensiometers can also be connected to pressure transducers and logged electronically. The tensiometer can only be used at relatively low tensions, between about 0 and "1.0 atm, but this range is important within the context of plant growth; plants obtain water via their roots by exerting a tensional force of about 0.05-15 atm.

Temperature

Soil temperature is an extremely dynamic property, varying diurnally and seasonally, with the effects being most rapid and extreme towards the surface. On a diurnal time-scale, soils are heated during the day and the effect gradually extends downwards, perhaps taking several hours to reach a depth of 30 cm. At night soils cool rapidly at the surface and heat is transferred upwards from within the soil. Seasonal heating and cooling cycles operate in a similar manner, but they penetrate deeper into the soil than diurnal cycles because the time-scale is much greater; diurnal cycles usually affect only the upper 30 cm or so, whereas seasonal cycles can penetrate to a depth of several metres. In order to detect their rapid variation, soil temperatures must be measured frequently and at a variety of depths, and are usually recorded electronically using a series of thermistors linked to an automatic recorder. Soil temperature is influenced by a number of soil properties, in particular texture, moisture and organic content; these will be discussed later in the context of soil influences on small-scale climatic characteristics.

Mechanics

Commonly encountered soil mechanical properties are strength, stability and consistence. The mechanical strength and stability of soils are derived from interparticle and interped forces responsible for the development of soil structure. Aggregate stability is a useful measure of the structural stability of soils, and is based on the ability of aggregates to survive wetting. During wetting, air is trapped in pores within the peds, and the pressure of the trapped air increases until it is sufficient to cause aggregate breakdown or *slaking*. In addition, water entering the aggregates interferes with electrostatic forces of interparticle attraction and may dissolve cementing agents, thus causing further weakening. Measurements of aggregate stability have included determining the proportion by weight of aggregates retained on a 2 mm sieve after wet-sieving or after ultrasonic dispersion, although such methods have been questioned because they subject the soil to artificial forces for relatively short periods of time. The strength and stability of a soil can also be assessed from its resistance to compression and shear, which provides a useful indication of the degree of cohesion. Resistance to compression can be determined using a penetrometer, whilst a shear vane can be used to measure resistance to shear. These characteristics can also be established from a triaxial shear test, in which a soil core is first subjected to compression in order to determine its load-bearing capacity, and then to a shear test.

Soil strength is closely related to a number of soil properties, particularly texture, organic content, bulk density and moisture content. Coarse-textured soils frequently possess a relatively high degree of strength due to the often irregular nature of particle boundaries and the resulting large surface area of contact. Coarse-textured soils with smooth particle boundaries are considerably weaker, however, due to the smaller surface area of contact. Fine-textured soils are often strongly cohesive and well aggregated, and are particularly strong when dry.

When wet, however, the cohesive forces tend to break down and soil strength decreases dramatically. The presence of organic matter improves the cohesive properties of soils and increases their strength. Similarly, soil strength increases with increasing bulk density, due largely to the reduction in total porosity.

In contrast to cohesion, dispersion may reduce the strength and stability of soils. This commonly occurs in soils whose colloidal fraction is dominated by adsorbed sodium, as often occurs in semi-arid environments. Sodium ions are only loosely held by colloidal particles, and are unable to neutralise fully the surface negative charge. This situation promotes interparticle repulsion rather than attraction, thus resulting in breakdown of soil structure. Structure may also be disrupted by repeated shrinkage and swelling, which can be particularly active in soils containing large amounts of swelling clays.

Soil consistence describes the change in state of soil, from solid through plastic to liquid, with increasing moisture content. It is established from the three Atterberg limits—shrinkage, plastic and liquid limits. The shrinkage limit represents the minimum moisture content above which soil volume begins to increase with further wetting; below this limit soil volume remains constant irrespective of changes in moisture content. The plastic limit represents the minimum moisture content above which the soil becomes mouldable, and the liquid limit represents the minimum moisture content above which the soil flows under its own weight. The plastic and liquid limits in particular are closely associated with a number of soil properties, for example cation exchange capacity, clay content and specific surface area.

Colour

Soil colour is a characteristic which can easily be determined in the field, and which provides useful information regarding the presence or absence of certain soil constituents. For example, dark colours are usually indicative of high organic contents, while red colours are characteristic of soil rich in iron oxides, and blue-grey colours indicate the presence of iron in its reduced form. Such indications are, however, only general because, for example, manganese can also produce a dark colouration, while the intensity of colour is often related to moisture content. The identification of soil colour is also rather subjective, as different individuals interpret colour in different ways. In an attempt to reduce subjectivity, soil colour is most often determined using the Munsell system where soil samples are matched against standard colour charts. Munsell colour notation has three components—the *hue* which indicates the major colour(s) present, the *value* which is a measure of the degree of darkness or lightness of the colour, and the *chroma* which is a measure of colour intensity. Pages of the chart are shown according to hue, and each colour chip on the page has its own co-ordinate, expressed in terms of the value on the vertical axis and the chroma on the horizontal axis. A sample of soil is therefore represented by a notation according to the colour to which it most closely corresponds, and each notation has an equivalent description; for example, a soil with a hue of 10YR, a value of 3 and a chroma of 4 is represented as 10YR 3/4, which has the description yellowish brown. The objective determination of soil colour has been the subject of much discussion, and although the Munsell system is the most widely adopted, other methods and indices have been developed.

SOIL CHEMICAL PROPERTIES

Elements and Compounds

Elements and compounds in a soil occur in two principal forms—as the chemicals that make up the structure of the basic soil constituents, and as individual components which are held in the soil by interparticle attraction. The first of these are important in terms of soil properties.only once the constituents start to break down, whereupon they are released into the soil, but the second of these are generally of more immediate importance because they are more readily available for interaction. The chemicals that make up the structure of mineral material are determined by total chemical analysis, for example by atomic absorption spectrometry following dissolution of the material in strong acids, or by the more rapid method of X-ray fluorescence spectrometry. In terms of organic matter, the principal structural elements which are normally determined are carbon and nitrogen. Organic carbon content is usually determined by dichromate oxidation, and this can also be used to estimate the organic matter content of a soil by using a conversion factor of 1.7 (organic matter per cent = organic carbon per cent × 1.7). Organic matter content can also be determined by loss on ignition or by digestion with hydrogen peroxide, although these methods may not always be particularly accurate. Nitrogen content is most often determined by the Kjeldahl digestion procedure.

Individual elements and compounds which are commonly examined as individual components include exchangeable bases, free iron and aluminium, carbonates and heavy metals. Exchangeable bases are held in the ion exchange complex of a soil and are usually extracted using ammonium acetate, then analysed by flame emission or atomic absorption spectrometric methods. Exchangeable cation content is commonly expressed in units of milliequivalents (me) per 100 g of dry soil, which represents the amount of a cation that will replace or combine with 1 mg of hydrogen per 100 g of dry soil.

Exchangeable base content can also be expressed in units of cmol(+)kg, which is exactly the same as me per 100 g. Free iron and aluminium are present in soils in a number of forms, including sesquioxides (oxide and hydroxide compounds), poorly crystalline or amorphous allophanic and imogolitic materials, and organometallic complexes. These can be determined using a range of selective extractants. For example, total free iron and aluminium is often determined by extraction with sodium dithionite or by a dithionite-citrate-bicarbonate (DCB) extraction. Inorganic forms of iron and aluminium, which are poorly crystalline or amorphous, can be extracted by sodium oxalate, and organically bound forms by potassium pyrophosphate.

The potency of these extractions decreases in the order: DCB> oxalate> pyrophosphate, and there is thought to be relatively little overlap between them. It is therefore possible to determine levels of all these different forms by sequential extraction performed on the same sample. The most common carbonates in soils are compounds of calcium and magnesium, derived largely from carbonate-rich parent materials.

Presence or absence of carbonates can be established simply by adding a few drops of dilute hydrochloric acid (HCl) to a small sample of soil; effervescence is observed if carbonates are present. More accurate measurement can be made in the laboratory, for example using a calcimeter, which measures the calcium carbonate ($CaCO_3$) equivalent by determining the quantity of carbon dioxide (CO_2) evolved during treatment with hydrochloric acid.

Heavy metals can occur naturally in soils, but often occur at enhanced levels due to additions from motor vehicle emissions, sewage sludge applications, metal mining and smelting, and scrap metal processing. They can exist in a number of forms including numerous compounds, particularly oxides, sulphides and sulphates, metal cations such as lead, zinc and cadmium (Pb^{2+}, Zn^{2+} and Cd^{2+}), which may be adsorbed onto the surfaces of negatively charged colloidal particles, and organo-metallic complexes. Heavy metals are commonly extracted from soils by acid digestion, using a strong mineral acid such as nitric acid to determine total contents, and EDTA or a weak organic acid such as acetic acid to determine 'plant-available' levels. Contents are then measured by atomic absorption spectrometry.

Ion Exchange

Ion exchange is a most important soil property in that it plays a key role in plant nutrition, and in a broader context, in the development of many chemical characteristics of soils. Central to ion exchange is the way in which ions are held on the surfaces of colloidal particles. Colloidal material consists mainly of clay and humus particles and is often referred to as the *clay-humus complex* or *exchange complex*. These particles have a high specific surface area (ratio of surface area to volume) and possess both high surface energy and significant surface charge. This charge is largely negative and can occur either as permanent charge or variable (pH dependent) charge. In the case of permanent charge, cations in mineral structures are replaced by cations of similar size but with lower valency (isomorphous substitution); such charge is independent of acidity. In the case of variable charge, hydrogen ions undergo reversible dissociation from surface groups such as "OH and "OH^+ on the edges of clay minerals and oxides, and "COOH, "OH and "NH_2 in organic material. As a result of their negative surface charge, colloidal particles behave like giant anions and are known as *micelles*. The micelles are able to attract cations onto their surfaces, a process known as *adsorption*. With increasing distance from the micelle surface, the concentration of cations decreases exponentially whilst the concentration of anions shows a reciprocal increase. The zone of adsorbed cations, together with the surface negative charge on the micelle, are often referred to as the *electrical double layer*.

The ease with which a cation is adsorbed depends on its valency and degree of hydration. Cations with a high valency have a high energy of adsorption and are therefore adsorbed in preference to lower valency cations. For cations with equal valency, however, the one with the smallest *radius of hydration* (radius of the ion together with the associated water molecules surrounding it) is adsorbed preferentially because it can get closer to the micelle surface. The generally accepted sequence of preferential adsorption for base cations is Ca^{2+}> Mg^{2+}> K^+> Na^+, and this is reflected in their usual proportions in the soil

(Ca^{2+}=80 per cent, Mg^{2+}=15 per cent, K^{+}+Na^{+} =5 per cent). Adsorbed cations can be exchanged for cations in the soil solution.

The sequence of preferential adsorption again applies, with cations of low valency or high radius of hydration being exchanged in preference for cations of higher valency or smaller radius of hydration, depending on their availability. Plants obtain many of their nutrients through this exchange mechanism, the nutrient cations being taken up through the root system in exchange for hydrogen ions.

Although cation adsorption and exchange tend to predominate in soils, in some circumstances anion adsorption and exchange are more common. While the former occur in soil containing significant amounts of humus, and clay minerals with a high specific surface area such as montmorillonite, vermiculite and illite, the latter occur in soil with limited humus content and with clay minerals with a low specific surface area such as kaolinite. Oxides of iron and aluminium are also significant in anion adsorption as they possess variable surface charge which is positive under acidic conditions. This positive charge results from the combination of hydrogen (H^{+}) ions with edge hydroxyl groups.

As in the case of cations, anions with a high valency are able to get closer to surfaces than anions with a low valency. Thus sulphate (SO_4^{2-}) and phosphate (PO_4^{3-}) are often strongly held, whereas nitrate (NO_3) tends to be held rather weakly. Like cations, adsorbed anions can also be exchanged for anions in the soil solution.

Unlike Ca^{2+}, Mg^{2+} and K^{+}, which are important plant nutrients, Na^{+} is toxic to many plant species, and it also has a deleterious effect on soil structure, promoting the dispersal of aggregates. The assessment of soil salinity can therefore be important in these instances, and various expressions are commonly used. The *sodium adsorption ratio* (SAR) represents the amount of Na^{+} relative to Ca^{2+} and Mg^{2+}.

The SAR is approximately equal to the *exchangeable sodium percentage* (ESP) of the soil, where the Na^+ level, measured in me per 100 g, is expressed as a percentage of the total exchangeable base content. An ESP value of 15 per cent is generally accepted as the threshold above which soils are considered to be sodium-affected. Soil salinity is usually determined from a saturation extract, where distilled water is added to the soil to produce a paste. Salinity is then established from the electrical conductivity of the saturation extract.

The ability of soil to yield cations is measured by its *cation exchange capacity* or CEC. In effect it is a measure of the amount of negative surface charge, and hence of the potential for cation adsorption. CEC is often determined by ammonium acetate extraction, with cations on the soil exchange complex being displaced by ammonium ions, although this method can be unreliable due to the change in pH, induced by ammonium acetate treatment, producing a change in surface charge; alternative extraction methods have therefore been developed, but these have not yet become widely used. As with exchangeable base content, CEC is expressed in units of me per 100 g or cmol(+)kg. Values vary dramatically and tend to be highest in soils with high clay and organic contents. CEC values for organic matter may be 150-300 me per 100 g, virtually twice that of clay. Clay mineralogy also has a major influence on CEC, in terms of the charge density per unit area; CEC values range from 5-10 me per 100 g for kaolinite, through 30-40 for illite, to > 100 for vermiculite and montmorillonite.

Acidity and pH

Acids in aqueous solutions undergo dissociation to release their constituent ions, namely hydrogen (H^+) and an acid anion.

Acidity is measured in terms of the H^+ ion concentration using the pH scale. The relationship between pH and H^+ ion concentration is inverse and logarithmic:

The pH scale ranges from 1.0 at the most acidic extreme to 14.0 at the alkaline extreme, with a value of 7.0 at neutrality.

The pH of soils varies widely, from around 2.0 in acid sulphate soils to about 12.0 in alkaline sodic soils; good quality agricultural soils have a value around 6.0 to 7.0.

The measurement of soil pH is usually made in a standard suspension of 1:2.5 weight to volume, in order to ensure data comparability. Although distilled water is often used to make up the suspension, a suspension made with a dilute solution of calcium chloride is sometimes used in order to provide a more realistic value of H^+ concentration by minimising calcium release from the soil exchange complex. For this reason pH levels measured in calcium chloride suspension are generally lower than those recorded in a suspension made up with distilled water. The measurement itself can be made using electrometric or colourimetric techniques.

In addition to H^+ ions, Al^{3+} ions play an important role in the generation of soil acidity, particularly in soils that are already acidic. Al^{3+} ions undergo hydrolysis during which H^+ ions are released into the soil solution.

Positively charged hydroxy-aluminium species can then occupy exchange sites, thus resulting in reduced CEC. Hydroxy-aluminium species may undergo further hydrolysis to produce yet more H^+ ions and stable aluminium hydroxide (gibbsite):

Consequently, soil acidity promotes the development of further acidity through aluminium hydrolysis, and this becomes an important source of H^+ ions when soils become acidic. If soil pH falls below about 5.5, Al^{3+} ions themselves begin to occupy exchange sites. Because of their higher valency, Al^{3+} ions are adsorbed much more strongly than divalent and monovalent cations, therefore levels of exchangeable aluminium increase, and amounts of exchangeable bases decrease, as pH declines. Soil acidity is closely related to many other soil properties such as organic content, exchangeable base content and CEC.

Aeration

Soil aeration relates to the amount of oxygen present in the soil atmosphere. This can be assessed using both direct and indirect methods. The rate of oxygen diffusion through the soil can be measured directly using a platinum electrode, while the concentration of oxygen in a sample of soil air can be determined by gas-liquid chromatography. A number of instruments are available for sampling the soil atmosphere and for monitoring changes in aeration, although problems have been experienced in their design, due largely to leakage. Indirect assessment of soil aeration can be made from detection of fermentation and putrefaction products, either by quantitative analysis or simply by their odour. Limited aeration can also be seen by the appearance of reduced compounds of manganese and iron, which are characterised by the presence of black specks and pallid grey colours respectively.

A particularly useful indicator of the degree of soil aeration is the *redox potential* (Eh) or oxidation-reduction status. Redox or oxidation-reduction reactions are those in which a chemical species undergoes oxidation or reduction through the transfer of electrons (e^-). An example of such a reaction is the reduction of ferric (Fe III) hydroxide to ferrous (Fe II) hydroxide.

This reaction is reversible, with a tendency to change towards the left under oxidising conditions and towards the right under reducing conditions. The redox potential (Eh) is the difference in electrical potential between the two halves of this coupled reaction. It can be measured using a platinum electrode and an appropriate reference electrode; these are inserted into the soil and the electrical potential difference (V) is recorded, and referred to a standard pH value of 7.0. Eh values in aerobic soils are generally between 0.3 and 0.8 V, while in anaerobic soils they are often between 0.3 and "0.4 V.

Under aerobic conditions, electrons produced during respiration combine with oxygen. Under anaerobic conditions, however, oxygen is unavailable and other chemical species must act as electron receptors. Ferric iron compounds often take on this role, undergoing reduction to ferrous iron compounds. Each chemical species has a threshold redox potential below which it begins to act as an electron receptor and thus becomes unstable. Ferrous iron compounds are often present in the centre of peds where Eh may be 0.1 V lower than at the edges, whereas ferric compounds are more likely to occur between peds, in areas adjacent to root channels, and in patches where texture is relatively coarse. Such microscale variation in distribution of the different iron compounds can often give the soil a mottled appearance, where blue-grey colours of ferrous compounds are interspersed with orange-brown colours of ferric compounds. The degree of soil aeration is therefore closely associated with porosity and pore size distribution, and is inversely related to soil water content. It is also related to texture and the degree of structural development. Optimum levels of aeration are most often found in well structured soils with a fine loamy texture.

Chapter 4

Soil Fertility

ELEMENTS ESSENTIAL FOR PLANT GROWTH

Higher plants require 16 nutrient elements to complete their life cycle. Of these, 13 are extracted from the soil and can be found in organic or inorganic fertilizers. The 13 essential nutrients are thought of as three distinct groups—primary macronutrients, secondary macronutrients, and micronutrients. Primary macronutrients are the elements needed in the largest quantity; these are nitrogen (N), phosphorus (P), and potassium (K). Secondary macronutrients are needed in lower concentrations and are calcium (Ca), magnesium (Mg), and sulfur (S). Micronutrients are needed in even lower concentrations and include iron (Fe), manganese (Mn), boron (B), chlorine (Cl), zinc (Zn), copper (Cu), and molybdenum (Mo).

Table: Essential Elements Required by Plants, Their Chemical Symbol, the Form Taken Up by the Plant, and Their Concentrations in the Plant.

Element	Chemical Symbol	Form Taken Up	Concentration
Carbon	C	CO_2	
Hydrogen	H	H_2O	
Oxygen	O	H_2O	
	Primary Macronutrients		
Nitrogen	N	NH_4^+, NO_3^-	1 to 5per cent

Phosphorus	P	$H_2PO_4^-$, HPO_4^{2-}	0.1 to 0.4per cent
Potassium	K	K^+	1 to 2per cent
		Secondary Macronutrients	
Calcium	Ca	Ca^{2+}	0.5per cent
Magnesium	Mg	Mg^{2+}	0.2per cent
Sulfur	S	SO_4^{2-}	0.15 to 0.2per cent
		Micronutrients	
Iron	Fe	Fe^{2+}, Fe^{3+}	50 to 200 ppm
Manganese	Mn	Mn^{2+}	20 ppm
Boron	B	H_3BO_3	10 to 50 ppm
Chlorine	Cl	Cl^-	100 ppm
Zinc	Zn	Zn^{2+}	20 to 50 ppm
Copper	Cu	Cu^{2+}	20 ppm
Molybdenum	Mo	MoO_4^{2-}	0.1 to 0.2 ppm

The carbon (C), hydrogen (H), and oxygen (O) utilized by a plant come from carbon dioxide and water. Little can be done to control the availability of these three except through drainage, irrigation, and modification of the physical condition of the soil. On a dry-matter basis, C, H, and O make up more than 94per cent of the plant biomass. This means the remaining 6per cent of the biomass is made up of the other 13 nutrients. Even though their amounts seem small, deficiency of only one essential element can limit the growth potential of a plant.

Removal of the primary and secondary macronutrients (N, P, K, Ca, Mg, and S) by various crops is reported in Table. The values listed in the table indicate average nutrient removal and represent only those nutrients found in the harvested portion of the crop. The values reported are not the quantities of nutrients needed to generate the crop yields shown. Keep in mind crop nutrient content can vary widely under different growing conditions, and soil nutrient availability is determined by various fixation and release mechanisms.

Fixation and release mechanisms that control soil nutrient availability are strongly influenced by soil pH. Figure 3-1 shows the relative availability of 12 essential nutrients at different pH levels for mineral soils. Figure 3-2 shows the relative availability of the same elements at different pH levels for organic soils. Chlorine is not reported because it is usually present in sufficient amounts.

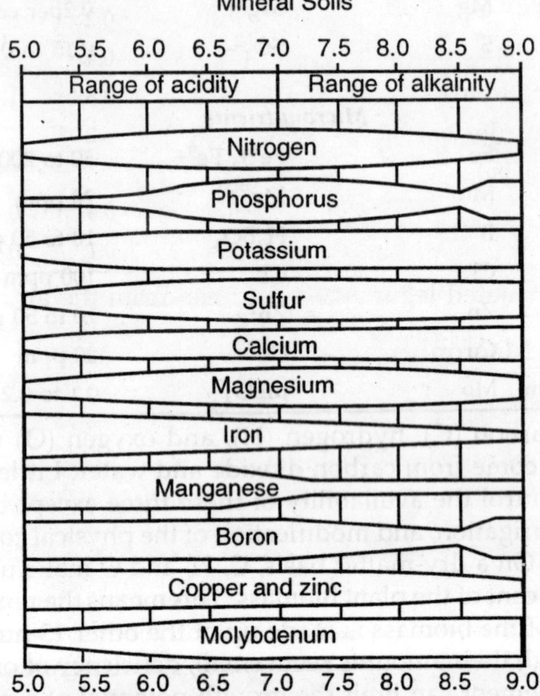

Fig. Relative availability of elements essential to plant growth at different pH levels for mineral soils.

Table: Approximate Amounts of Primary Macronutrients Removed by Various Crops.

Crop (Removal Units)	N	P_2O_5	K_2O
Alfalfa (lb/ton)	57*	13	50
Corn (lb/bu)			
Grain	0.9	0.4	0.3
Stover	0.7	0.2	1.1
Corn-silage (lb/ton)	9.4	3.2	8.0
Cool-season grasses (lb/ton)	40	13	0

Oats (lb/bu)			
Grain	0.7	0.3	0.2
Straw	0.4	0.2	1.0
Sorghum-grain (lb/bu)			
Grain	0.8	0.2	0.2
Stover	0.6	0.4	1.7
Soybean (lb/bu)	3.8*	0.8	1.4
Sugarbeets (lb/ton)	4	2	5
Wheat (lb/bu)			
Grain	1.3	0.6	0.4
Straw	0.4	0.1	0.7

* Inoculated legumes fix nitrogen from the air.

Primary Macronutrients

Nitrogen (N)

Most nitrogen (N) contained in soil is in the organic form. One percent organic matter represents approximately 1,000 lbs nitrogen per acre in the top six inches of soil. Despite the abundance of nitrogen in organic matter, this form is unavailable to the crop. Organic nitrogen must be mineralized (converted into ammonium) by soil microbes to become plant available. The rate of mineralization is controlled by soil pH, moisture, temperature, and aeration and is highly variable. Crops grown on organic soils (greater than 20per cent organic matter) typically require less fertilizer nitrogen than crops grown on mineral soils due to mineralization of organic nitrogen. Ammonium present in a warm, well-aerated soil is quickly converted (one to two weeks) to nitrate. Soil nitrate is highly mobile and is susceptible to both denitrification (finer soils) and leaching (coarser soils) losses.

Fertilizer nitrogen sources can be classified into three categories—inorganic, synthetic organic, and natural organic. There is little difference between sources of nitrogen when properly applied at equivalent rates. Typical fertilizer nitrogen sources and their analysis are presented in Table. To assure maximum agronomic and economic production, nitrogen should be managed such that losses are minimized. Best management practices that decrease the potential for nitrogen loss should be selected—such as subsurface injection or dribble banding of liquid nitrogen rather than broadcast application, incorporation of surface-applied urea (conventional till) or application coinciding with a rainfall event (no-till), and spring application of nitrogen for summer crops rather than late fall application.

Table: Analysis and Classes of Some Nitrogen (N) Fertilizer Materials.

Nitrogen Fertilizer	Formula	Form	Percent Nitrogen
Inorganic			
Ammonium nitrate	NH_4NO_3	Solid	34
Ammonium sulfate	$(NH_4)_2SO_4$	Solid	21
Anhydrous ammonia	NH_3	Gas	82
Aqua ammonia	NH_4OH	Liquid	20
Nitrogen solutions	Varies	Liquid	20 to 32
Synthetic			
Urea	$CO(NH_2)_2$	Solid	46
Natural organic			
Animal manure		Solid	1 to 20
Sewage Sludge		Solid	5 to 10

1 Liquid under pressure.

If nitrogen is to be applied in the fall for a spring crop, anhydrous ammonia should be selected and should only be applied if the soil temperature is below 50°F (other sources of nitrogen for fall application are not promoted). There is a risk, however, that warm, wet conditions may prevail in the spring that can lead to significant losses of nitrogen, requiring later additions of nitrogen to maximize production. Spring-applied nitrogen should be applied as close to planting as possible but be aware that application of anhydrous ammonia followed closely by planting can result in decreased germination. Allow some time to pass between anhydrous application and planting. Application of nitrogen at sidedress is typically the most efficient method.

Urea applied in no-till production systems should coincide with an expected rainfall event. Volatilization losses of urea fertilizers without adequate incorporation (by tillage or rainfall) can be significant. Liquid nitrogen formulations should preferably be injected below the soil surface or dribbled in a band to minimize volatilization losses. Rates of application should be based on realistic yield goals determined from historical production levels. Specific rates of application are available in the Tri-State Fertilizer Recommendations for Corn, Soybeans, Wheat, and Alfalfa.

Phosphorus (P - P_2O_5)

Soils can contain between 500 to 1,500 pounds of total phosphorus (P) per acre, most of which is unavailable to the crop. The level of available phosphorus is quite variable across the state, and field levels of phosphorus can be determined by soil test. Historical management and soil pH (of mineral soils) determine how much phosphorus is plant available. Phosphorus is taken up by the plant in two primary forms—$H_2PO_4^-$ and HPO_4^{2-}. As soils become more acidic, $H_2PO_4^-$ is the primary species taken up, but excessive soil acidity can result in phosphorus deficiency as aluminum and iron concentrations increase, "fixing" phosphorus. Conversely, as soils become more basic or alkaline, HPO_4^{2-} is the primary species taken up, but high pH can result in phosphorus deficiency as calcium phosphate precipitates. Plants take up both phosphorus forms indiscriminately, so there is no specific pH level that results in maximum phosphorus availability. In general, soil pH should be maintained between 6.0 to 7.5 to maximize plant available phosphorus.

Phosphorus fractions of typical phosphorus fertilizers are water soluble, citrate soluble, citrate insoluble, and total. Citrate soluble phosphorus is referred to as "available" phosphorus, and the portion of citrate soluble phosphorus that dissolves in water is water-soluble phosphorus. Phosphorus content of phosphorus fertilizers is expressed as percent citrate soluble phosphorus as P_2O_5-. Most commercial fertilizers contain 50per cent or greater water-soluble phosphorus and are suitable for row application. Typical phosphorus fertilizers and their analysis are presented in Table.

Table: Analysis of Phosphate (P_2O_5) and Potassium (K_2O) Fertilizers.

Fertilizer Material	Formulma		Total	Available	Water Soluble	
			Approximate per cent			
		N	P_2O_5			K_2O
Phosphorus Carriers						
Concentrated superphosphates	$Ca(H_2PO_4)_2$	0	47	45	85	0
Monoammonium phosphate	$NH_4H_2PO_4$	11	49	48	92	0
Diammonium phosphate	$(NH_4)_2HPO_4$	18	47	46	90	0
Ammonium polyphosphate	$[NH_4PO_3]n$	10	34	34	100	0
Potassium Sources						
Potash	KCl	0	0	0	0	62
Potassium Nitrate	KNO_3	13	0	0	0	44
Potassium sulfate	K_2SO_4	0	0	0	0	50
Potassium magnesium sulfate	$K_2SO_4 \bullet 2\ MgSO_4$	0	0	0	0	21

Despite the fact that plants take up the ortho form of phosphorus, poly-based forms of phosphorus are just as effective at satisfying plant needs. Poly forms of phosphorus added to the soil are actually converted to ortho forms relatively quickly.

Potassium (K – K_2O)

Soils contain between 10,000 to 20,000 ppm potassium. Despite this high amount, only a small portion is actually plant available. Exchangeable (adsorbed to soil CEC—Cation Exchange Capacity) and solution potassium make up the plant-available portion. Potassium availability is dependent upon soil mineralogy and rainfall. Soils that develop from minerals high in potassium (feldspars and micas) have naturally high potassium fertility levels. Soils that developed under high rainfall conditions can be quite deficient in potassium because it has been leached out of the soil.

Muriate of potash is the primary source of potassium fertilizer used commercially. It is readily soluble and contains over 62per cent potassium.

Secondary Macronutrients

Calcium (Ca)

Calcium is one of the most abundant nutrient elements found in the soil and is rarely deficient. Calcium availability is strongly tied to soil pH, and soils that are maintained at adequate pH (>5.5) levels should have adequate calcium. If calcium is deficient, addition of lime to increase soil pH will remedy the problem. The ultimate source of calcium is lime (calcitic or dolomitic—both contain relatively high calcium amounts). Sources of calcium and their analysis are shown in Table.

Table: Analysis of Some Ca, Mg, and S Fertilizers

Material	**Average Percent**					
	N	*P*	*K*	*Ca*	*Mg*	*S*
Ammonium nitrate limestone	21			7	4	23
Magnesium sulfate (Epsom salt)					10	13
Calcium sulfate (gypsum)				22		17
Sul-po-mag			22		11	23
Superphosphate						
Normal		18-20		20		12-14
Concentrated		30-50		13		1
Ammonium sulfate	21					24
Potassium sulfate			48-51		17-18	
Elemental sulfur						50-99

Ammonium phosphate	11	48	2
Ammonium phosphate (sulfate)	16	20	15
Manganese sulfate			14-17
Sulfuric acid			33

1 Liming materials also contain varying levels of Ca and Mg.

Magnesium (Mg)

Magnesium deficiencies, while rare, can occur, primarily in the eastern half of the state. Like Ca, magnesium availability is strongly tied to soil pH. Soils with neutral or basic pH should have adequate magnesium. If soil magnesium and pH are low, use of dolomitic lime to neutralize soil acidity will remedy the problem. If magnesium is low and the pH is near neutral, application of one-half to one ton of 12per cent magnesium (dolomitic) lime will provide enough magnesium for maximum plant production. This will not result in over-liming of medium- or fine-textured soils.

Some propose an optimum calcium to magnesium ratio to provide proper nutrition for growing plants. There is, however, no economic benefit to this methodology. Depending upon soil mineralogy and soil pH, applying enough calcium or magnesium to reach an ideal ratio can be extremely expensive. Several research projects across the Corn Belt, have revealed that maximum production levels can be attained at several different Ca:Mg ratios.

Sulfur (S)

Sulfur deficiencies are rare but can be found in forage production systems, especially on sandier soils low in organic matter. Much like nitrogen, the primary form of sulfur in the soil is found in the organic fraction, and the form taken up by higher plants is highly mobile. For every 1per cent of organic matter, there are approximately 140 lbs of sulfur, which like nitrogen, must be mineralized to be plant available. Sulfur is deposited in large quantities from rainfall primarily due to industrial activities. As emission standards decrease sulfur release from industrial processes, S fertilization may become more important. Fertilizer sulfur is available from many different sources which are reported in Table.

Micronutrients

Micronutrient levels across the state are adequate for maximum plant production, and deficiencies are rare. Specific field environments and soil conditions increase the potential of finding a micronutrient deficiency. If a micronutrient is found to be deficient, remember that over-application of most micronutrients can result in toxicity. Plant tissue analysis is the best way to determine if a plant has a micronutrient deficiency.

Molybdenum (Mo) and boron (B) can reach toxic levels even when applied in small quantities. Recommendations should be followed closely when either of these elements is being applied. Boron should not be applied in the row for corn or soybean. Manganese (Mn) toxicity occurs on many soils in eastern region when pH nears 6.0. Alfalfa and soybeans are especially sensitive to excess manganese. Foliar-applied manganese in excess of recommended amounts, or in small quantities of water, may burn leaves of wheat, oats, and sugarbeets.

Micronutrient deficiencies are strongly influenced by soil pH. Soils that are acidic (< 6.5) should have adequate levels of most of the micronutrient metals (Fe, Mn, Zn, B, Cu). Molybdenum behaves just the opposite of the other micronutrients; its availability increases as soil pH increases.

Animal Manure

Animal manure is a good source of plant nutrients and contains many of the elements essential for plant growth. It is especially important to farming operations that include livestock enterprises. It provides those operators the opportunity to utilize efficiently the waste produced. Soils should be tested prior to applying manure. Table shows the average nutrient content (N, P, K) of three different manure sources. Because the nutrient value of manure is closely related to the dietary regimen, which can result in vastly different nutrient levels, manure should be tested to determine its nutrient value prior to land application.

Table: Average and Range of Nutrient Concentrations in Manures (Units Are Percent of Dry Matter).

Type of Manure	N (per cent)	P_2O_5 (per cent)	K_2O (per cent)
Dairy cattle	4.3	1.8	4.0
	(2.2–14.3)	(0.7–5.2)	(1.2–14.8)
Swine (finishing)	14.0	4.3	5.3
	(2.7–24.2)	1.4–6.8)	(0.4–16.3)
Poultry (layers)	3.9	7.0	3.8
	(2.3–5.6)	(1.6–17.9)	(1.6–6.7)

Manure should be analyzed for total solids, total nitrogen, ammonium nitrogen, phosphorus, potassium, calcium, and magnesium. Other nutrient analyses should be available upon request.

Calculating Fertilizer Rates

Once the analysis of a specific fertilizer material is known (typically displayed as a percent), rates of application can be easily computed using this equation:

Pounds of fertilizer material needed = nutrient rate (lbs per acre) ÷ percent analysis of material

For example, if the soil test calls for 150 lbs of Mg and a dolomitic liming material is used that is 12per cent Mg, the rate of lime required is 1,250 lbs per acre (150/0.12). This calculation works for all fertilizer materials.

LIME AND LIMING MATERIALS

Proper use of both lime and fertilizer is necessary for high crop yields. To optimize production, soil acidity should be corrected prior to fertilizer application.

Liming benefits soil in the following ways:

- Supplies calcium and magnesium.
- Increases soil pH and the availability of phosphorus, molybdenum, and magnesium.
- Reduces harmful concentrations of aluminum, manganese, and iron.
- Increases favorable microbial activity, which results in an increased release of organically bound nitrogen, phosphorus, sulfur, and other nutrients.

Determination of Lime Requirement

Soil pH measures the active soil acidity or alkalinity. The lime requirement is determined using the buffer pH, or the lime test index, which measures potential soil acidity. Finer-textured soils that have relativelyhigh CEC (Cation Exchange Capacity) have the ability to buffer changes in soil pH by releasing adsorbed hydrogen; thus, more lime is required to affect soil pH. Coarser-textured soils have lower CEC values and have a diminished ability to provide hydrogen to replace that which is neutralized by liming; thus, less lime is required. The lower the buffer pH is below 6.8, the greater the lime requirement. Table shows the relationship between the buffer pH and lime requirement to various pH levels. For organic soils, pH is used to determine lime requirement.

Liming Materials

Agricultural liming materials used for correcting soil acidity include all calcium and magnesium oxides, hydroxides, carbonates, silicates, or combinations sold for agricultural purposes.

Liming materials are labeled based on their effective neutralizing power (ENP). The ENP of a liming material considers the material equivalence, purity, fineness of grind, and percent moisture. Particle size of liming materials impacts their effectiveness at neutralizing soil acidity and their speed of reaction. Ag liming materials typically contain particles of differing sizes which results in longer-term acid neutralization.

Smaller particles react quicker while larger particles dissolve slowly, affecting soil pH over a longer period. This is why liming is typically not necessary every year. When comparing liming materials and their associated cost, ENP provides a good way to identify the most economical source.

Lime Recommendations

Lime recommendations presented in Table are in tons of material/acre and assume a liming material with an ENP of 2000.

Recommendations should be adjusted if:

- Another grade of liming material is to be used.
- Depth of plowing is different from that indicated (eight inches).
- A different crop is planted.

Adjustments for the Type of Liming Material

Although the activity of liming materials in the soil is the same, liming materials do differ. Liming materials differ in their purity, fineness, and moisture, which affect the rates of application necessary to alter soil pH.

The example given here assumes a three-ton-per-acre lime requirement. If a liming material that has an ENP of 2,000 lbs/ton is used, simply apply three tons of that material per acre. Assume that a liming material that has an ENP of 1,500 lbs per ton is chosen. To compute the new recommendation, divide the lime requirement by the quotient of the ENP divided by 2,000:

3.0 tons per acre ÷ (1,500 lbs/ton ÷ 2,000) = 4 tons per acre

If a liming material with an ENP of 1,500 lbs/ton is used to meet a lime requirement of three tons per acre, 4.0 tons of the liming material should be applied per acre.

Adjust for the Depth of Tillage

If plowing depth is different from eight inches, the lime requirement must be adjusted. To adjust the lime requirement, simply use a multiplier to change the lime requirement. Multipliers for differing depths of tillage are presented in Table. For example, assume that the lime requirement is four tons per acre, but the depth of tillage is only six inches. Multiply four tons per acre by 0.75:

4 tons per acre * 0.75 = 3.0 tons per acre

It only requires three tons of lime per acre to achieve the desired change in soil pH to a depth of six inches.

Table: Adjustments in Liming Rate for Depth of Tillage.

Tillage Depth (inches)	Multiplying Factor
3	0.38
6	0.75
7	0.88
Base 8	1.00
9	1.13
10	1.25
11	1.38
12	1.5

No-Till Adjustments

Lime adjustments for no-till cultural practices are also based on tillage depth. Two soil samples are recommended for no-till row crop cultural practices—shallow sample (zero- to four-inch depth) and deep sample (zero- to eight-inch depth). The shallow sample should be submitted to determine the lime requirement. If lime is required to neutralize surface acidity, decrease the lime requirement from Table by half. The deep lime requirement remains the same. When lime is applied to the surface, it should be lightly incorporated. If the slope of the field is steep enough to cause erosion, do not incorporate at all. When lime is not incorporated, do not use urea fertilizer for a year after lime application. Ammonium nitrate, anhydrous ammonia, or banded 28per cent solution are suitable N materials for this case.

Other Adjustments in Lime Recommendations

A minimum lime requirement of two tons per acre is recommended for forage legumes if the soil pH is 6.2 or less, even if the buffer pH is higher than 6.8 and lime would typically not be recommended.

Sandy soils are so weakly buffered that despite the soil pH being below optimal ranges, the buffer pH value is greater than 6.8. A one ton per acre lime requirement should be used if soil pH is 0.3 pH units below the desired soil pH. A two ton per acre lime requirement should be used if soil pH is 0.6 pH units below the desired soil pH.

Why Re-liming Is Necessary

Periodic liming is necessary because basic cations (Ca, Mg, and K) are removed by crops and can be lost by leaching and erosion. Ammoniacal forms of nitrogen fertilizers also contribute to soil acidity. Coarser-textured soils require more frequent applications of lime than finer-textured soils.

Acidic Subsoils

Most of the soils in eastern region and the light-coloured soils of western region have acidic subsoils. If there is any question about the pH of the subsoil, the subsoil should be sampled and tested separately.

When lime applications are necessary to correct subsoil acidity, the lime requirement to increase pH to 6.8 should be used. The pH level in the topsoil should be 6.8 or above to promote downward movement of the lime. Only where the surface pH is maintained near 6.5 will the subsoil pH increase. Because this downward movement takes several years, the sooner the lime is applied, the better.

Lime recommendations of more than four tons per acre should be applied as a split application to achieve a more thorough mixing of the lime with the soil. Half the lime should be applied prior to tillage and half before subsequent tillage. For best results, lime should be applied at least six months before seeding. This allows the lime time to neutralize soil acidity. When a maintenance application is recommended, it can be applied any time in the cropping sequence.

Organic Soils (Muck and/or Peat)

Organic soils usually do not benefit from liming unless the pH of the soil in the root zone is below 5.3. If the pH of the surface is near 5.3, but the subsurface pH is 4.0 or below, it may be necessary to lime and deep plow. A favorable pH to a minimum depth of two feet should be maintained to accommodate root penetration. In general, lime does not move downward further than plow depth in an organic soil. The pH of organic soils should be determined on samples taken to depths of 24 inches.

High Organic Matter Soils

Usually the most desirable pH range for organic soil is 5.3 to 5.8. When the organic matter content of soils is between 10per cent and 20per cent, the pH should be shifted as indicated in Table.

Table: The Desired pH for High-Organic-Matter Mineral Soils.

Percentage of Organic Matter	Desired pH
10	6.0
15	5.8
20 or higher	5.3 to 5.8

DIAGNOSTIC METHODS

Soil Testing

Soil testing provides information about the nutrient level of the soil and the amounts of lime and fertilizer necessary to maximize production.

Typical soil analysis includes:

- Soil pH and buffer pH
- Available phosphorus
- Exchangeable potassium

- Cation Exchange Capacity (CEC)
- Exchangeable calcium
- Exchangeable magnesium

Additional soil analysis can include:

- Organic matter
- Nitrate-nitrogen
- Soluble salts
- Total heavy metals
- Available manganese, zinc, and boron

Most soil testing labs report soil nutrient levels in parts per million (ppm). The Tri-State Fertilizer Recommendations for Corn, Soybean, Wheat, and Alfalfa uses both ppm and pounds per acre (lb per acre) to identify critical nutrient concentrations. Make certain that similar units are being compared. To convert from ppm to lb per acre, simply multiply ppm by 2.

Soil Sampling

When sampling a field, randomly collect 20 to 25 soil cores to a depth of eight inches. Thoroughly mix the samples together to form a composite and submit a sample of the composite to a testing lab in an approved container. Soil samples should be collected either in the fall after harvest or in the spring. Soil samples should be collected about the same time each year to avoid extreme changes in soil-test information. If lime application is necessary, samples should be collected in the fall. Soil samples should not be taken during the growing season. Soil samples should be collected every two to three years depending upon soil conditions. Coarser-textured soils should be tested more frequently than finer-textured soils.

Soil pH and Buffer Ph

Soil pH measures the active acidity/alkalinity of the soil solution. Buffer pH (may be reported as LTI-lime test index, which is buffer pH multiplied by 10) provides a measure of the active and potential soil acidity, and determines the lime requirement. Coarser-textured soils typically have higher buffer pH values than finer-textured soils. As buffer pH decreases, the lime requirement increases.

Available Phosphorus

Soils typically contain several hundred pounds of phosphorus per acre. A great majority of this soil phosphorus is not plant available. To determine the amount of phosphorus that is and will become plant available, an acidic solution is mixed with the soil that dissolves aluminum phosphate precipitates. Available phosphorus of acid soils in the North Central Region is measured using the Bray-Kurtz-P-1 procedure.

Exchangeable Potassium

Much like soil phosphorus, soil potassium (K) is present in large quantities. Only a small fraction of this total amount is plant available. Unlike phosphorus, potassium is not precipitated with other ions in the soil solution, but is adsorbed on the CEC of the soil or trapped between soil mineral layers. To measure exchangeable potassium, a solution containing a high concentration of another cation is added to the soil (usually ammonium acetate). The ammonium from the solution knocks potassium off the CEC and other exchange sites. The amount knocked off is measured and used to identify how much will become plant available over the growing season.

Cation Exchange Capacity

Cation Exchange Capacity (CEC) measures the capacity of a soil to adsorb cations, including hydrogen (H^+), calcium (Ca^{2+}), magnesium (Mg^{2+}), and potassium (K^+). Other cations including aluminum (Al^{3+}) and iron (Fe^{3+}) are also adsorbed, but in slightly acidic to neutral soil, their amounts are small enough to be ignored. In slightly acidic to neutral soils, calcium and magnesium take up approximately 80per cent of the CEC, while potassium only occupies less than 5per cent. In acidic soils, aluminum and hydrogen can begin to occupy a larger percentage of the CEC.

To determine the CEC of a soil, use this equation:

CEC = ppm Ca/200 + ppm Mg/121 + ppm K/390 + 1.2 * (70 – [buffer pH * 10])

The CEC of a soil depends largely on the soil texture and the amount of organic matter present. The larger the CEC value, the more cations the soil is capable of adsorbing, which decreases leaching. Attempts to increase the CEC of a soil by adding clay or organic matter are impractical due to the amounts that would be necessary to affect a change. Liming acidic soils only affects the CEC slightly.

The normal range in CEC for different soil textures is as follows:

Soil Textures	Common CEC Range (meq/100 g soil)
Coarse (sands)	1 to 5
Medium (silts)	6 to 20
Fine (clays)	21 to 30
Organic soils	30 plus

Percent base saturation of the soil CEC usually falls within the following ranges:

Element	Range in Percent Saturation*
Calcium	40 to 80
Magnesium	10 to 40
Potassium	1 to 5

*Assuming the pH value is in the recommended range.

Calcium to Magnesium Ratio

The calcium to magnesium ratio is based on their percent saturation of the CEC. If the ratio of calcium to magnesium is 1:1 or less (less calcium than magnesium) and the soil calls for application of lime, calcitic (low magnesium, high calcium) lime should be applied. Most plants grow well across a wide range of calcium to magnesium ratios.

Magnesium to Potassium Ratio

To avoid potential grass tetany problems, the magnesium to potassium ratio should stay above 2 to 1. In other words, the percent magnesium saturation of the CEC should be twice that of the percent potassium saturation. As more potassium is taken up, less magnesium is absorbed, resulting in a nutrient imbalance in the forage. This imbalance can be transferred to the grazing animal in the form of grass tetany.

Soluble Salts

The soluble salts test indicates the concentration of all fertilizer and non-fertilizer salts in the soil. Excessive salt levels, known as saline soil conditions, can be toxic to plants, especially germinating or young plants. Saline soil impairs the ability of the plant to extract soil water, leading to a drought-like symptom. Obviously, the drier the soil gets, the more apparent the problem is. Saline spots in the field are typically characterized by good tilth and excessive moisture retention. Severe brine solution spills cause excessive soluble salt concentrations. To reduce the salt concentration, the soil should be well drained and leached with high-quality water. Natural rainfall gradually reduces the soluble salt level in most well-drained. Table provides a guide for interpreting soluble salt levels.

Table: Soil and Plant Conditions for Various Soluble Salt Concentrations.

Soil and Plant Condition	Soluble Salt Concentration (mmhos/cm)
Unfertilized, leached field soils	0.15
Well-fertilized soil for optimum plant growth	1-2
Growth of salt-sensitive crops affected	Greater than 2
Severe injury to plants	Greater than 3

Values in table should be reduced by one-half for prolonged droughty conditions. Toxicity of a single salt is greater than an equivalent amount of a mixture of salts.

Plant Analysis

Plant analysis is not a substitute for soil testing, but rather supplements soil testing by determining whether nutrient deficiencies are occurring. Plant analysis can be used to diagnose suspected nutrient deficiencies or detect nutrient deficiencies before plant growth is limited. Plant analysis also contributes to more economical and efficient use of fertilizer materials, avoiding excessive or inadequate application rates.

Used in conjunction with other data and observations, a plant analysis report aids in evaluating the nutrient elements in the soil-plant system. It also provides a way to evaluate the effectiveness of fertilizers added to the soil. Plant analysis can also help determine response to fertilizer treatment by answering the question: Was the nutrient element supplied by the fertilizer sufficiently absorbed by the plant? Plant analysis is especially useful for determining whether the soil is adequately supplying required micronutrients.

Plant Sampling

Each crop has its own sampling methodology and sampling techniques for the major agronomic crops as shown in Table. When sampling young plants (seedlings), collect the above-ground portion of 10 to 20 plants. Plant samples can be analyzed for all major nutrients (this does impact the cost of analysis). The desired concentration of a nutrient should occur within the sufficiency range for that nutrient. Table lists the typical sufficiency ranges of nutrients for corn, soybean, alfalfa, and small grains. Nutrient concentrations slightly lower than shown indicate a marginal condition, which may adversely affect plant growth. The limits for these ranges vary depending on crop, plant part, and state of growth when sampled. These values relate specifically to a particular plant part sampled at a specific stage of growth.

When sampling plants in an obviously stressed area, it may be beneficial to submit a check tissue sample which is from an adjacent area that is stress free. This helps further determine if a nutrient deficiency exists. The relative concentration of elements also helps determine sufficiency or deficiency and should be considered when interpreting plant analysis. For example, the ratio of potassium to magnesium, zinc to phosphorus, and manganese to iron assists in diagnosing suspected magnesium, iron, manganese, and zinc deficiencies.

Chapter 5

Corn Production

Successful corn production requires an understanding of the various management practices and environmental conditions affecting crop performance. Planting date, seeding rates, hybrid selection, tillage, fertilization, and pest control all influence corn yield. A crop's response to a given cultural practice is often influenced by one or more other practices. The keys to developing a successful production system are to recognize and understand the types of interactions that occur among production factors, as well as various yield limiting factors, and to develop management systems that maximize the beneficial aspect of each interaction. Knowledge of corn growth and development is also essential to use cultural practices more efficiently to obtain higher yields and profits.

HOW CLIMATE AFFECTS CORN PRODUCTION

Temperature

Corn can survive brief exposures to adverse temperatures—low-end adverse temperatures being around 32°F and high-end temperatures being around 112°F. Growth decreases once temperatures dip to 41°F or exceed 95°F. Optimal temperatures for growth vary between day and night, as well as over the entire growing season. For example, optimal daytime temperatures range between 77°F and 91°F, and optimal nighttime temperatures range between 62°F and 74°F.

The optimal average temperatures for the entire crop growing season, however, range between 68°F and 73°F.

Even though corn seeds germinate and grow slowly at about 50°F, the first spring planting dates usually begin when the average air temperatures reach 55°F, and soil temperature at seed depth is more favorable for seedling growth. Poor germination resulting from below-normal temperatures is the greatest hazard of planting too early. The growing point of germinating seedlings remains below or near the soil surface and usually is not vulnerable to freeze damage until plants reach the five- to six-leaf collar stage. By this time, corn is about 10-inches tall, and the probability of freezing temperatures greatly decreases. The loss of leaves from frost generally does not seriously injure small plants, although such loss may delay plant development.

Temperatures less than 40°F reduce photosynthesis, even if the only symptom is a slight loss of leaf colour. Frost injury symptoms may appear on leaves even when nighttime temperatures do not fall below the mid 30s; radiational cooling can lower leaf temperatures to several degrees below air temperatures on a clear calm night. If frost kills leaves but not stalks before physiological maturity (black layer formation) in the fall, sugars usually continue to move from the stalk into the ear. However, yields are generally lower, and harvest moisture may be high because of high grain moisture at the time of frost and slow drying rates following premature death.

High temperature stress during ear formation, reproduction, and grainfill can reduce yield, but temperatures less than 100°F usually do not cause much injury if soil moisture is adequate. Under rain-fed conditions, corn usually begins to stress when air temperatures exceed 90°F during the tasseling-silking (pollination) and grainfill stages. Corn yield may be reduced 1.5 bushels per acre for each day the temperature reaches 95°F or higher during pollination and grainfill. Extended periods of hot, dry winds may cause tassel blasting and loss of pollen.

Precipitation

A corn crop typically uses 20 to 22 inches of water during the growing season. Water requirements of corn vary according to the stage of development, as shown in Table. Corn reaches its peak water use during pollination, when plants are silking.

Table: Water Use Rates for Corn at Different Growth Stages.

Growth Stage	Water Use Rate, Inches/Day
Prior to 12-leaf stage	< 0.20
12-leaf	0.24
Early tassel	0.28
Silking	0.30
Blister kernel	0.26
Milk	0.24
Dent	0.20
Full dent	0.18

Excessive rainfall, resulting in flooding and ponding of soils, may cause serious injury to a corn crop depending on its stage of development. The major stress caused by flooding and ponding is a lack of oxygen needed for the proper function of the root system. When plants are very small (prior to the six-leaf collar stage), they generally are killed after about five or six days of submersion. Death occurs more quickly (within two to four days) if the weather is hot, because warm temperatures speed up the biochemical processes that use oxygen, and warm water has less dissolved oxygen. Cool weather, on the other hand, may allow plants to live for more than a week under flooded conditions.

As soon as plants reach the six- to eight-leaf collar stage and the plant's growing point is above the soil surface, plants can tolerate a week or more of standing water—not necessarily without harm. In older plants, total submersion may increase disease incidence, and plants will suffer from reduced root growth and function for some days after the water recedes. Tolerance of flooding generally increases with plant age, but reduced root function resulting from a lack of oxygen is probably more detrimental to yield before and during pollination than during rapid vegetative growth or grainfill.

Nutrient uptake is also reduced in soils saturated by excessive rainfall. Not only does poor aeration inhibit effective root development and function, but the anaerobic conditions associated with saturated soils promote denitrification. Frequent rainfall can also cause nitrate leaching.

For crop moisture to be adequate, available soil moisture must be more than sufficient to meet the atmospheric evapourative demand. On windy, hot, sunny days with low humidity, evapouration demand on a crop is high, and a high amount of available soil moisture must be present if the crop is to avoid stress. Under cloudy skies, high humidity, and cooler temperatures, atmospheric evapourative demand is low, and plants can get by with lower amounts of available soil moisture.

The soil must provide a corn crop with enough water to offset the amounts lost through transpiration. If these needs are not met, the plant will wilt. Table shows the effect of drought on corn grain yield from four consecutive days of visible wilting. Through the late vegetative stage, corn is fairly tolerant of dry soils. Mild drought during June may even be beneficial because roots generally grow downward strongly as surface soils dry, and the crop benefits from the greater amount of sunlight that accompanies dry weather. From the two weeks before through the two weeks following pollination, corn is very sensitive to drought, however, and dry soils during this period may cause serious yield losses. Most of these losses result from pollination failure, and the most common cause is the failure of silks to emerge from the end of the ear. When this happens, the silks do not receive pollen; thus, the kernels are not fertilized and will not develop. Drought later in grainfill has a less serious effect on yield, though root function may decrease and kernels may abort or not fill completely.

Table: Effects of Drought on Corn Yield During Several Stages of Growth.

Stage of Development	Percent Yield Reduction
Early vegetative	5 to 10
Tassel emergence	10 to 25
Silk emergence, pollen shedding	40 to 50
Blister	30 to 40
Dough	20 to 30

* After four consecutive days of visible leaf wilting.

Source: Claassen, M. M., and R. H. Shaw. 1970. Water deficit effects on corn. II. Grain components. Agron. J. 62:652-655.

Drought stress often leads to plant nutrient stress. The shallow depths where fertilizer is placed are dry under drought situations, which may limit nutrient uptake.

CORN GROWTH AND DEVELOPMENT

Corn growers who understand how the corn plant responds to various cultural practices and environmental conditions at different stages of development are able to use management practices more efficiently and, thus, obtain higher yields and profits. Knowledge of growth and development may also help in troubleshooting problems related to abnormal growth caused by pest problems or inappropriate cultural practices.

Table describes two of the most widely used staging systems for corn development. Extension agronomists use the Leaf Collar Method throughout the United States; crop insurance adjusters, however, use the Horizontal Leaf Method to assess hail and other weather-related plant damage. Table shows a timeline relating corn growth and development to normal heat unit accumulation and calendar dates during the growing season.

Table: Growth Staging Systems for Corn.

	Vegetative Stages		Reproductive Stages
VE	Emergence	R1	Silking
V1	First leaf collar	R2	Blister
V2	Second leaf collar	R3	Milk
V3	Third leaf collar	R4	Dough
V(n)	nth-node collar	R5	Dent
VT	Tasseling	R6	Physiological maturity

Horizontal Leaf Method

Growth staging system used by hail adjusters for hail damage assessment.

1. Identify uppermost leaf that is 40per cent to 50per cent exposed and whose tip is below the horizontal.
2. Typically, a "horizontal leaf" growth stage will be 1 to 2 leaf stages greater than the collar method.

Although this table was developed to determine yield losses resulting from hail damage, it can also be used to help assess losses resulting from other defoliation injuries (such as wind, insect feeding, herbicide damage, and foliar N "burn"). The most common damage from hail is loss of leaf area, although stalk breakage and bruising of the stalk and ear may also be severe. Note that the largest yield losses result from defoliation damage that occurs during the late vegetative stages and the reproductive stages (silking and tasseling). Defoliation at early growth stages does not affect yield the same way as it does at later growth stages because much of the plant's total leaf area is not yet exposed. Extensive defoliation of plants in the 10-leaf growth stage (or V8, eight leaf collar stage) does not result in a large yield loss because only 25per cent of the leaf area is exposed, and the plant can easily recover from early damage. On the other hand, severe damage to plants during tasseling results in a large yield loss because, by that time, 100per cent of the leaf area has been exposed and cannot be replaced.

Table: Effect of Corn Leaf Area Loss at Various Growth Stages.*

Growth Stage**	Percent Leaf Area Destroyed					
	10	20	40	60	80	100
	Percent Yield Loss					
7 leaf	0	0	0	4	6	9
10 leaf	0	0	4	8	11	16
13 leaf	0	1	6	13	22	34
16 leaf	1	3	11	23	40	61
Tasseled	3	7	21	42	68	100
Late milk	1	3	10	21	35	50
Dent	0	0	3	10	17	24

* Adapted from Corn Loss Instructions, NCIS Publication No. 6102, Rev. 1984.

** Based on horizontal leaf method for staging plant growth (two-leaf stages greater than leaf collar method).

Early killing frost in the fall may damage immature corn and reduce yield. The effect of frost damage to corn depends on the severity of defoliation, stalk damage, and stage of growth.

Hybrid Selection

Selecting a group of hybrids for planting is a key step in designing a successful corn production system. To stay competitive, growers must introduce new hybrids to their acreage on a regular basis. During the past 40 years, the genetics of corn hybrids has improved steadily, which has contributed to steady increases in grain yield potential ranging from 0.7per cent to 2.6per cent per year.

Growers should choose hybrids best suited to their farm operations. Corn acreage, soil type, tillage practices, desired harvest moisture, and pest problems determine the need for such traits as drydown rate, disease resistance, early plant vigour, plant height, and more. End uses of corn should also be considered. Will the corn be used for grain or silage? Will it be sold directly to the elevator as shelled grain or used on the farm? Capacity to harvest, dry, and store grain should also be considered.

Maturity

Growers should choose hybrids with maturity ranges appropriate for their geographic area or circumstances. Corn for grain should reach physiological maturity or "black layer" (maximum kernel dry weight) one to two weeks before the first killing frost in the fall. Use days-to-maturity and growing-degree-day (GDD) ratings along with grain moisture data from performance trials to determine differences in hybrid maturity. Although yields of full-season hybrids usually exceed those of short-season hybrids, early- to mid-maturing hybrids have been developed in recent years with yield potential approaching those of full-season types. Late- to full-season hybrids do not always mature or dry down adequately before frost, which results in wet grain. When confronted with delayed planting or replanting decisions, growers may need to switch to early- to medium-maturity hybrids adapted to their area, but they should avoid short-season hybrids that are earlier than those normally used in their area.

Days-to-Maturity Rating System

The most common maturity rating system is the days-to-maturity system. This system does not reflect actual calendar time between planting and maturity—106-day hybrid, for example, does not actually mature 106 days after planting. A days-to-maturity rating is based on relative differences within a group of hybrids for grain moisture at harvest. A one "day" maturity difference between two hybrids is typically equal to a one-half to three-fourths percentage point difference in grain moisture. For example, a 106-day hybrid would be, on average, 3 to 4.5 points drier than a fuller season 112-day hybrid if they were planted the same day (6 "days" multiplied by 0.5 or 0.75).

The relationship between days to maturity and kernel moisture is usually dependable when comparing hybrid maturities within a single seed company. However, because there are no industry standards for the days-to-maturity rating system, grain moisture comparisons of similar hybrid maturities from different seed companies may vary considerably. Days-to-maturity ratings are satisfactory for pre-season hybrid maturity selection when length of the growing season is usually not an issue. For delayed planting or replanting hybrid selection needs, growers need more absolute descriptions of a hybrid's growing season requirements to manage the risk of a killing fall frost to late-planted corn.

Growing-Degree-Day (GDD) Maturity Rating System

The growing-degree-day (GDD) maturity rating system is based on heat units. It is more accurate in determining hybrid maturity than the days-to-maturity system because growth of the corn plant is directly related to the accumulation of heat over time rather than the number of calendar days from planting. The GDD system has several advantages over the days-to-maturity system. The GDD system provides information for choosing hybrids that will mature reliably, given a location and planting date; allows the grower to follow the progress of the crop through the growing season; and aids in planning harvest schedules.

The GDD calculation method most commonly used for corn in the United States is the 86/50 cutoff method. Growing degree days are calculated as the average daily temperature minus 50.

$$GDD = (Tmax + Tmin \div 2) - 50$$

If the maximum daily temperature (Tmax) is greater than 86°F, 86 is used to determine the daily average. Similarly, if the minimum daily temperature (Tmin) is less than 50°F, 50 is used to determine the daily average. The high cutoff temperature (86°F) is used because growth rates of corn do not increase above 86°F. Growth at the low temperature cutoff (50°F) is already near zero, so it does not continue to slow as temperatures drop further. Growing degree days are calculated daily and summed over time to define thermal time for a given period of time. The cumulative GDDs associated with different vegetative and reproductive stages.

As with any system, the GDD system has several shortcomings. Growing-degree-days ratings of hybrids with similar days-to-maturity ratings don't always agree, especially if the hybrids are from different companies. Some seed companies start counting GDDs from the day of planting, while others begin from the day of emergence. When this occurs, similar maturity hybrids may vary by 100 to 150 GDDs—the average GDDs required for emergence. Some companies use entirely different mathematical methods to calculate GDDs. Although most companies use the "86/50 cutoff method" described earlier, others use different methods to calculate GDDs. Also, under certain delayed planting situations and stress conditions, GDD requirements for maturity may be reduced significantly.

Yield Potential and Stability

Choose hybrids that have produced consistently high yields across a number of locations and/or years. The Corn Performance Tests indicate that hybrids of similar maturity may vary in yield potential by as much as 40 bushels per acre or more. Choosing a hybrid because it possesses a particular trait, such as big ears, many kernel rows, deep kernels, prolificacy, or upright leaves, does not ensure high yields; instead, look for stability in performance across environments.

The superior performance of corn hybrids has virtually eliminated the use of open-pollinated varieties. The average yield for the open-pollinated varieties was 83 bushels per acre and 29per cent lodged stalks. The average yield of the hybrids was 133 bushels per acre with 8per cent lodging—a 50-bushel or 60-percent yield advantage for hybrids over open-pollinated varieties.

Growers may want to consider planting herbicide-resistant or -tolerant hybrids in soils where residual levels of certain herbicides may be too high for normal corn growth. Seed companies have used various genetic techniques to insert resistance or tolerance to imazethapyr plus imazapyr (Lightning), glufosinate ammonium (Liberty), and glyphosate

(Roundup) into existing corn hybrids which are characterized as Clearfield (or imidazolinone-tolerant), Liberty Link, and Roundup Ready hybrids, respectively. Planting these tolerant/herbicide-resistant hybrids may allow growers to use herbicide formulations normally used on soybeans or other crops. These herbicide resistant/tolerant hybrids offer new weed control options that involve fewer applications and use of more environmentally benign chemicals.

Corn hybrids highly resistant to European corn borer injury and western corn rootworm also are available. These resistant hybrids contain a gene from bacteria that produces the insecticide known as Bt. Planting Bt corn hybrids may eliminate the need for soil insecticide treatments (rootworm) and postemergent insecticide applications (corn borer), which are less effective and potentially harmful to nontarget beneficial insects.

Review the results of state, company, and county performance trials before purchasing hybrids. Because weather conditions are unpredictable, the most reliable way to select superior hybrids is to consider performance during the last year and the previous year over a wide range of locations and climatic conditions.

On-farm strip tests are not reliable in hybrid selection because they cannot predict hybrid performance across a range of environmental conditions. However, on-farm hybrid tests can be useful in evaluating various traits, such as lodging, drydown, harvestability, disease resistance, and staygreen.

Stalk Quality and Lodging

Hybrids with poor stalk quality should be avoided for grain production even if they show outstanding yield potential. This trait is particularly important in areas where stalk rots are perennial problems or where field drying is anticipated—conditions that often lead to lodging (stalk breakage below the ear following crop maturation). If growers have their own drying facilities and are prepared to harvest at relatively high moisture levels (above 25per cent) or are producing corn for silage, then standability and fast drydown rates are less critical selection criteria.

Traits associated with improved hybrid standability include resistance to stalk rot and leaf blights, genetic stalk strength, short plant height and ear placement, and high staygreen potential. Staygreen refers to a hybrid's potential to stay healthy late into the growing season, after reaching maturity, and should not be confused with late maturity. Resistance to European corn borer conferred by the Bt trait can also enhance stalk quality by limiting entry points in plant tissue through which fungal pathogens can invade the plant. However, the Bt trait will do little to minimize stalk rot and lodging in a hybrid characterized by below-average stalk quality.

Another stalk-related problem, green snap or brittle snap, has started to appear in recent years. Corn plants are more prone to snapping during the rapid elongation stage of growth. According to studies in Iowa, Minnesota, and Nebraska, the V5 to V8 stages (corn approximately 10 to 24 inches in height) and the V12 stage through tasselling are the most vulnerable stages. Vulnerability to green snap damage does vary among hybrids. However, all hybrids are at risk from wind injury when they are growing rapidly prior to tasseling. The use of growth regulator herbicides, such as 2,4-D or Banvel, has also been associated with stalk brittleness, especially if late application or application during hot, humid conditions occurs. Once tassels begin shedding pollen, green snap problems generally disappear.

Disease Resistance and Tolerance

Hybrids should be selected for resistance or tolerance to stalk rots, foliar diseases, and ear rots, particularly those that have occurred locally. Seed dealers should provide information on hybrid reactions to specific diseases.

Grain Quality

The protein and oil composition of corn grain is a major factor in determining grain feeding value. Although the grain market does not include this factor in price determination, growers who feed livestock may use this information to reduce feed costs and optimize diets. Because hybrid genetics significantly affect the protein and oil content of corn grain, there is interest in using compositional data in selection of hybrids to plant for feed use, as well as other end uses. For feed, protein content is of primary interest, whereas for processing uses, oil content is of interest. Corn grain is typically 8per cent protein and 3.6per cent oil (on a 15.5per cent moisture basis).

Table shows averages and ranges for the grain protein and oil content of hybrids tested at different locations. The protein and oil content of grain is reported as protein and oil percentages at 15.5per cent grain moisture. Although significant differences among hybrids for oil and protein were found in each test, protein and oil levels varied considerably from test to test. Some normal dent corn hybrids produced primarily for grain exhibit elevated protein and oil levels. Environmental conditions (temperatures, rainfall) and cultural practices (nitrogen fertility, plant population) can affect grain composition, especially grain protein.

Table: Average Protein and Oil Content of Corn Grain (at 15.5per cent Grain Moisture).

Year	Region	Early Maturity Test Protein	Early Maturity Test Oil	Full Season Test Protein	Full Season Test Oil
		per cent	per cent		
2001	SW	7.5	3.3	7.4	3.5
	NW	8.5	3.7	8.5	3.9
	NE	8.9	3.8	8.4	3.8
2002	SW	8.5	4.0	8.5	4.1
	NW	8.9	4.3	9.1	4.4
	NE	8.1	3.9	8.4	4.1
2003	SW	8.5	3.8	8.4	3.8
	NW	8.0	3.7	8.0	3.8
	NE	8.3	3.6	7.9	3.7

Date of Planting

Planting should begin before the optimum date if soil conditions will allow the preparation of a good seedbed. Growers should have the equipment capability to plant more than half of their corn acres prior to the optimum planting date; this should allow planting all the corn acres prior to the calendar date when corn yields begin to quickly decline. Corn producers usually cannot perform field operations during all days of their optimum planting date range due to spring rains and cool weather conditions that limit soil drying. On average, during the optimal corn planting time, only one out of three days is available during which fieldwork can occur.

Table shows the effect of date of planting in Columbus. In central, yields decline approximately 1 to 1.5 bu/day for planting delayed beyond the first week of May. Grain yield and test weight were increased by early plantings, whereas grain moisture was reduced, thereby allowing earlier harvest and reducing drying costs. Early planting generally produces shorter plants with better standability. Delayed planting increases the risk of frost damage to corn and may subject the crop to greater injury from various late insect and disease pest problems, such as European corn borer and gray leaf spot.

Corn should be planted only when soils are dry enough to support traffic without causing soil compaction. The yield reductions resulting from mudding the seed in may be much greater than those resulting from a slight planting delay. No-tillage corn can be planted at the same time as conventional, if soil conditions permit. In reality, however, planting may need to be delayed several days to permit extra soil drying. Planting a full-season hybrid first, then alternately planting early-season and mid-season hybrids, allows the grower to take full advantage of maturity ranges and gives the late-season hybrids the benefit of maximum heat unit accumulation. Full-season hybrids generally show greater yield reduction when planting is delayed compared with short- to mid-season hybrids. Planting early hybrids first, followed by mid-season, and finally the full-season hybrids spreads the pollination interval for all the corn acres over a longer time period and may be a good strategy for some drought-prone areas with longer growing seasons.

Planting hybrids of different maturities reduces damage from diseases and environmental stress at different growth stages (improving the odds of successful pollination) and spreads out harvest time and workload. Consider spreading hybrid maturity selections between early-, mid-, and full-season hybrids—for example, a 25-50-25 maturity planting, with 25per cent in early- to mid-season, 50per cent in mid- to full-season, and 25per cent in full-season. Planting a range of hybrid maturities is probably the simplest and most effective way to diversify and broaden hybrid genetic backgrounds.

When corn planting is delayed past the optimum dates or if a crop needs to be replanted, it may be necessary to switch hybrid maturities. In most delayed planting situations, however, full-season hybrids still perform satisfactorily and reach physiological maturity (black layer formation) when planted as late as the last week of May. Hybrids planted in late May or early June mature at a faster thermal rate (require fewer heat units) than the same hybrid planted in late April or early May.

Other factors concerning hybrid maturity need to be considered when planting is delayed. For plantings in late May or later, the dry-down characteristics of hybrids should be considered. Although a full-season hybrid may still have some yield advantage over shorter season hybrids planted in late May, the full-season hybrid could have significantly higher grain moisture at maturity than earlier maturing hybrids. In addition, there will be less calendar time for field drying, and drying costs will be higher. Later planting dates generally increase the possibility of damage from European corn borer (ECB) and may warrant selection of ECB Bt hybrids.

Seeding Depth

The appropriate planting depth varies with soil and weather conditions. For normal conditions plant corn 1.5 to two-inches deep to provide frost protection and allow for adequate root development. Shallower planting often results in poor root development and should be avoided in all tillage systems. In April, when the soil is usually moist and the evapouration rate is low, seed should be planted shallower—no deeper than 1.5 inches. As the season progresses and evapouration rates increase, deeper planting may be advisable. When soils are warm and dry, corn may be seeded more deeply—up to two inches on non-crusting soils. Seed press wheels can help ensure good seed-soil contact, which is especially important as temperatures increase to 70°F or 80°F.

Row Width

This reduction in row spacing coincided with an increase in average plant population from approximately 18,000 plants/A to more than 26,000 plants/A. During the past decade, there has been considerable interest in narrowing row spacing even further, and many studies have been performed comparing corn planted in narrow rows and conventional 30-inch row spacing.

Although narrow row systems are often perceived as a proven method for increasing yield and profitability, recent studies on narrow-row corn production have produced mixed results. Some of the inconsistency may be related to latitude with narrow rows in the north central region of the United States exhibiting the largest yield increases (5per cent or more) over 30-inch rows. This advantage diminishes sharply moving southward with average yield advantages for narrow rows of less than 2per cent in the central Corn Belt and less further south. Results of a Michigan State University study conducted in 1998-99 showed that corn grain yields increased by 2per cent and 4per cent when row width was narrowed from 30 inches to 22 inches and 15 inches, respectively.

Some growers are considering "twin rows" as another row spacing configuration that may offer some of the yield increases associated with narrow row corn. In the typical twin row system, two rows are placed 7-inches apart on 30-inch centers, although other twin row configurations are used. Twin rows make it possible to create narrow rows without changing the row configuration of other equipment, and to avoid costs associated with equipment conversion to a narrow row system. Staying on 30-inch centers allows growers to use the same corn header and tractor tire spacing used in 30-inch corn production.

Potential yield gains from narrow rows must be balanced against the investment for new equipment and higher input costs associated with narrowing row spacing. Greater interest in increasing equipment use efficiency by using the same planter or drill for soybean, sugar beet, and corn may warrant adoption of narrow row systems for corn. Producers in northern regions that also grow soybeans and sugar beets in 22-inch rows often find it more efficient to use this same row spacing for corn.

Plant Populations and Seeding Rates

When corn is produced for grain, recommended plant populations at harvest (or final stand) can range from 20,000 to 30,000+ plants/A, depending on the hybrid and production environment. Hybrids differ in their response to high plant population with some exhibiting stalk lodging at the upper end of the plant population range. Populations for corn silage typically exceed those for grain by 2,000 to 4,000 plants/A. Seed companies specify a range in final stands for the various corn hybrids they market, and these seeding rate guidelines should be followed closely to maximize crop performance.

Yield potential of the production environment is the primary factor determining hybrid yield response to increasing plant population. Seeding rate adjustments should be made on a field-by-field basis using the average yield potential of a site over a three- to five-year period as the major criterion for determining the appropriate plant population. When determining the realistic yield potential for a site over a five-year period, it may be appropriate to ignore the highest and lowest yields, which may have occurred during years that were unusually favorable or unfavorable for corn performance.

Higher seeding rates are recommended for sites with high yield potential that have high soil fertility levels and water holding capacity. On very productive soils which may average yields of 175 bu/A or more (such as a drained, Kokomo silty clay loam), final stands of 30,000 plants/A or more may be required to maximize yields. On soils averaging 150 bu/A, final stands of 26,000 to 28,000 plants/A may be needed to optimize yield. Lower seeding rates are preferable when droughty soils or late planting (after June 1) limit yield potential. On soils that average 120 bu/A or less, final stands of 22,000 plants/A may be adequate for optimal yields.

Hybrid response to high population can be limited by stalk lodging, which often increases at higher plant density. Some hybrids that have shown positive yield response to higher populations cannot be grown at high plant densities because of the increased risk of lodging at harvest. Lodging reduces yields and slows the harvest operation. Therefore, it is essential that hybrids planted at high seeding rates possess superior stalk quality for standability. Hybrids should also have resistance (or the best levels of tolerance available) to fungal leaf diseases (such as gray leaf spot and northern corn leaf blight), which contribute to stalk lodging problems and stalk rots (such as Anthracnose and Gibberella).

If a grower plans to rely extensively on field drying that can delay harvest, there may be little benefit from using high plant populations above 30,000 plants/A. We recently completed a study that evaluated effects of plant population (24,000 to 42,000 plants/A) and harvest dates (early-mid October, November, and December) on the agronomic performance of four hybrids differing in maturity and stalk quality.

Table: Harvest Date and Plant Population Effects on Grain Yield, Moisture, and Stalk Lodging.

Harvest Date	**Harvest Population (plants/acre)**			
	24,000	**30,000**	**36,000**	**42,000**
Yield, bu/acre				
Early/Mid Oct	191	194	197	198
Early/Mid Nov	187	194	193	188

Early/Mid Dec	172	174	167	161
Grain Moisture, per cent				
Early/Mid Oct	24.9	24.0	22.4	23.7
Early/Mid Nov	18.2	17.9	18.0	17.9
Early/Mid Dec	17.4	17.3	17.6	17.7
Stalk Lodging, per cent				
Early/Mid Oct	3	4	4	4
Early/Mid Nov	17	20	27	34.
Early/Mid Dec	33	42	52	59

Although the hybrids exhibited similar yield potential when harvested early (early/mid October), differences in yield became evident with harvest delays, which could be attributed to differences in stalk quality. Yield differences among plant population were generally small on the first harvest date, but with harvest delays, major yield losses occurred at the higher plant populations, especially 42,000 plants/A, due to increased stalk lodging. Grain moisture averaged about 24per cent on the first harvest date, 18per cent on the second harvest, and 17.5per cent on the third harvest date. After the first harvest in early/mid October, stalk lodging increased to as much as 80per cent for certain hybrids at high plant populations, resulting in yield losses of nearly 50per cent by mid December.

Final stands are always less than the number of seeds planted per acre. Cold, wet soil conditions, insects, diseases, cultivation, and other adversities will reduce germination and emergence. Generally, you can expect from 10 to 20 percent fewer plants at harvest than seeds planted. To compensate for these losses, you need to plant more seed than the desired population at harvest. Many seed companies recommend over-planting by 10per cent to 15per cent.

To calculate your own planting rate, consider the following formula:

Planting Rate = Desired Population per Acre ÷ (Germination × Expected Survival)

Germination is the percent seed germination shown on the seed tag (converted to decimal form). Expected survival is the percent of seedlings and plants that you expect to reach harvest maturity under normal conditions (converted to decimal form). Ninety percent survival (or 10per cent plant mortality) is about average. If you are planting very early when the soil will likely remain cool for several days following planting, you may want to reduce expected survival by 5per cent. A similar approach should be followed when planting no-till, especially in heavy residues.

Example:

Target stand at harvest - 26,000 plants per acre

Seed tag indicates 95per cent seed germination

Assume 90per cent survival (10per cent plant mortality)

Planting rate = 26000 ÷ (0.95 × 0.90) = 30,409 seedsper acre

According to the formula, you should consider a planting rate of approximately 30,400 seeds/A to achieve the desired final stand of 26,000 plants/A.

Uneven plant spacing and emergence may reduce yield potential. Seed should be spaced as uniformly as possible within the row to ensure maximum yields and optimal crop performance —regardless of plant population and planting date. Corn plants next to a gap in the row may produce a larger ear or additional ears (if the hybrid has a prolific tendency), compensating for missing plants. These plants, however, cannot make up for plants spaced so closely together in the row that they compete for sunlight, water, and nutrients. Crowding often results in barren plants or ears too small to be harvested (nubbins), as well as stalk lodging and ear disease problems.

Although uniformity of stand cannot be measured easily, studies have indicated that reduced plant stands will yield better if plants are spaced uniformly than if there are large gaps in the row. As a general guideline, yields are reduced an additional 5per cent if there are gaps of four to six feet in the row and an additional 2per cent for gaps of one to three feet. Studies at Purdue University suggest that corn growers could improve grain yield from four to 12 bushels per acre if within-row spacing were improved to the best possible uniformity (depending on the unevenness of the initial spacing variability).

The most effective way to improve planter accuracy is to keep planting speed within the range specified in the planter's manual. Additional considerations for improving seed placement uniformity are listed here:

- Match the seed grade with the planter plate.
- Check planters with finger pickups for wear on the back plate and brush (use a feeler gauge to check tension on the fingers, then tighten them correctly).
- Check for wear on double-disk openers and seed tubes.
- Make sure the sprocket settings on the planter transmission are correct.
- Check for worn chains, stiff chain links, and improper tire pressure.
- Make sure seed drop tubes are clean and clear of any obstructions.
- Clean seed tube sensors if a planter monitor is being used.
- Make sure coulters and disk openers are aligned.
- Match the air pressure to the weight of the seed being planted.

Uneven emergence affects crop performance because competition from larger, early emerging plants decreases the yield from smaller, later emerging plants. The primary causes of delayed seedling emergence in corn include soil moisture variability within the seed depth zone, poor seed-to-soil contact resulting from cloddy soils, inability of no-till coulters to slice cleanly through surface residues, worn disk openers, and maladjusted closing wheels. Other causes include soil temperature variability within the seed zone, soil crusting prior to emergence, occurrence of certain types of herbicide injury, and variable insect and/or soil-borne disease pressure.

Based on research at the University of Illinois and the University of Wisconsin, if the delay in emergence is less than two weeks, replanting increases yields less than 5per cent, regardless of the pattern of unevenness. However, if one-half or more of the plants in the stand emerge three weeks late or later, then replanting may increase yields up to 10per cent. To decide whether to replant in this situation, growers should compare the expected economic return of the increased yield with both their replanting costs and the risk of emergence problems with the replanted stand.

Making Replant Decisions

Although it is not unusual that 10per cent to 15per cent of planted seeds fail to establish healthy plants, additional stand losses resulting from insects, frost, hail, flooding, or poor seedbed conditions may call for a decision on whether or not to replant a field. The first rule in such a case is not to make a hasty decision. Corn plants can and often do outgrow leaf damage, especially when the growing point is protected beneath or at the soil surface (up until about the six-leaf collar stage). If new leaf growth appears within a few days after the injury, then the plant is likely to survive and produce normal yields.

When deciding whether to replant a field, assemble the following information: original planting date and plant stand, earliest possible replanting date and plant stand, and cost of seed and pest control for replanting. If the plant stand was not counted before damage occurred, providing that conditions for emergence were normal, estimate population by reducing the dropped seed rate by 10per cent. To estimate stand after injury, count the number of living plants in 1/1,000 of an acre. Take counts as needed to get a good average one count for every two to three acres.

When the necessary information on stands, planting, and replanting dates has been assembled, to locate the expected yield of the reduced plant stand by reading across from the original planting date to the plant stand after injury. Then, locate the expected replant yield by reading across from the expected replanting date to the stand that would be replanted. The difference between these numbers is the percentage yield increase (or decrease) to be expected from replanting.

Here's how these tables might be used to arrive at a replant decision. Let's assume that a farmer planted on May 9 at a seeding rate sufficient to attain a harvest population of 30,000 plants per acre. The farmer determined on May 28 that his stand was reduced to 15,000 plants per acre as a result of saturated soil conditions and ponding. The expected yield for the existing stand would be 79per cent of the optimum. If the corn crop was planted the next day on May 29 and produced a full stand of 30,000 plants per acre, the expected yield would be 81per cent of the optimum. The difference expected from replanting is 81 minus 79, or 2 percentage points. At a yield level of 150 bushels per acre, this increase would amount to three bushels per acre which would probably not justify replanting costs.

The effects of planting date and plant population on the final grain yield for the central Corn Belt. Grain yields for varying dates and populations in both tables are expressed as a percentage of the yield obtained at the optimum planting date and population.

Keep in mind that replanting itself does not guarantee the expected harvest population. Corn replant decisions early in the growing season will be based mainly on plant stand and plant distribution. Later in the season as yields begin to decline rapidly because of delayed planting, calendar date assumes increased importance.

FERTILITY RECOMMENDATIONS

A good nutrient management programme is key to high-yield corn production. Instituting the best management techniques to ensure adequate nutrient availability throughout the growing season can pay real dividends at the end of the year and minimize the adverse effects of nutrient runoff and leaching on the environment. Considering the importance of nitrogen (N), phosphorus (P), and potassium (K), the discussion will cover these primary soil fertility concerns.

Nitrogen

Timing and Sources

Nitrogen (N) applications for corn production should consider all risks and weigh all possibilities, particularly with nitrogen prices being high. Fall application of nitrogen is not recommended, but if N is to be applied in the fall, make certain that soil temperatures are below 50°F and that anhydrous ammonia is used. Do not apply N fertilizers that contain nitrate in the fall; the risk of loss is high due to leaching. Application of N in the spring is more efficient and less susceptible to loss. Nitrogen stabilizers may be used for early spring application, but the benefit of such compounds is questionable under certain growing conditions. Application of sidedress N is a good alternative to preplant applications of N. In-season applications move fertilization away from the busy planting period and are closer to actual crop uptake of N. Sidedressing also minimizes the risk of N loss, especially on poorly drained fine-textured soils, which are subject to denitrification, and coarse-textured soils, which are susceptible to leaching. The main risk of in-season application is the possibility of delayed application due to wet conditions.

When selecting a nitrogen source, remember that a pound of N is a pound of N; make selections based on risk and cost. For example, it would be risky to apply urea to the surface of no-till ground due to the potential loss of N by volatilization. Surface dribble-banding of liquid N or subsurface injection are better alternatives. This is not to say that urea is not a good source of N, but in this instance, there are better options. Always consider the cost of the material as well as the field environment that will be encountered to get the most efficient use of fertilizer N.

Rates

Current nitrogen recommendations for corn production based on yield potential are presented. Rates should be adjusted for high organic soils by using the pre-sidedress nitrate soil test or flat rate decreases of at least 40 lb per acre. Application rates for late planted corn should also be adjusted to reflect the decrease in yield potential. To determine a reasonable yield potential, consider the yield average of your last five corn crops at that specific field.

Phosphorus and Potassium

Application Methods

Phosphorus (P) and potassium (K) are a little easier to handle than N when it comes to application methods. Phosphorus and potassium are not subject to the same loss mechanisms as N, thus application concerns are not as restrictive. The main loss mechanism for P is runoff. Utilization of conservation practices that minimize the risk of soil runoff to surface waters is adequate for good P management. P and K can be applied either broadcast prior to planting or banded (near the row or over the row [pop-up]) as a starter when planting. If applying starter in a band two inches to the side and two inches below the seed, the total amount of salts applied (N + K2O) should not exceed 100 lb per acre. If starter is applied over the rows with the seed (not recommended due to potential salt problems), the total salts (N + K2O) applied should not exceed 5 lb per acre for low cation exchange capacity (CEC) soils or 8 lb per acre for high CEC soils. The benefit of starter fertilizers increases when soil-test levels and soil temperatures are low and when soil surface residues are high. Soils that have moderate to high levels of soil test P and K show little to no benefit from starter fertilizer.

Sources

Little difference exists between commonly used forms of phosphorus and potassium with regard to nutrient uptake. Ortho- and poly-phosphate formulations perform equally well, even though the crop takes up the ortho form (poly forms convert to ortho forms rapidly). It should be mentioned that if dry formulations of P are to be applied in contact with the seed, monoammonium phosphate (MAP) is a somewhat safer form of P to apply than diammonium phosphate (DAP). DAP produces more ammonia (NH_3) which is toxic to germinating seeds. When banding MAP, DAP, or ammonium polyphosphate (APP), do not exceed more than 40 lb N per acre. If soil-test P and K are high on no-till soils, then only N should be applied as a starter, unless 40 to 60 lb N per acre have been applied preplant.

Rates

Soil test levels below the critical value are considered deficient and warrant application of fertilizer. Buildup and maintenance recommendations are designed to increase soil-test levels to the critical value or maintain current soil-test levels. Considering it takes between 8 to 20 lbs of P_2O5 and 5 to 10 lbs of K_2O (added or removed) to change the soil test level by one unit (depends largely upon soil texture), soil-test levels above the critical value will be adequate for crop production for at least a few years.

Crop Rotations

The corn-soybean rotation is by far the most common cropping sequence used. This crop rotation offers several advantages over growing either crop continuously. Benefits to growing corn in rotation with soybeans include more weed control options, fewer difficult weed problems, less disease and insect buildup, and less nitrogen fertilizer use. Corn grown following soybeans typically yields about 10per cent more than continuous corn.

No-till cropping systems, which leave most of the prior crop residue on the surface, are more likely to succeed on poorly drained soils if corn follows soybean or meadow rather than if corn follows corn or a small grain, such as wheat. Table shows the influence of corn rotation on corn response to tillage and soil type. On the poorly drained Hoytville silty clay loam, where corn follows soybean or meadow, yield differences between no-till and plowed are greatly reduced. Crop rotation with soybeans had much less effect on corn response to tillage on the well-drained Wooster silt loam. This yield advantage to growing corn following soybean is often much more pronounced when drought occurs during the growing season.

CORN PEST MANAGEMENT

Weed Control

A number of factors need to be considered when developing weed control programmes for corn, including soil type, weeds, weeds present, crop rotation, and budget. No single control programme effectively handles the various weed problems that arise under different environmental conditions. Weeds are the major pest control problem in corn production.

Insect Control

Field corn is a host crop for a complex of pest species that may cause economic injury throughout the growing season. Despite the number of different pests that feed on corn, the average field generally exhibits minimal pest problems because the normal activity of most pest populations is too low to inflict significant injury to warrant attention. However, economic levels of pest injury do occur, and economic losses resulting from sudden outbreaks of pest activity can be prevented if the farm manager is aware of the biology of corn pests, monitors fluctuations in pest population activity, and implements timely corrective action.

Management of pest populations on corn requires a combination of preventive and timely responsive actions based on the risks associated with various cultural practices and pest activity observations collected through periodic field inspections. The text in the next section reviews the complex of invertebrate pests that may impact field corn, summarizes the ecological factors associated with outbreaks of specific pests, and discusses available pest management options.

Planting Time Decisions

When planting corn, there are basically four pest-management options:

- If potential loss from rootworm exists, then an application of a granular or liquid soil insecticide, the use of a seed-applied insecticide, or the use of a Bt hybrid for rootworm may be warranted.
- Application of an insecticide in a herbicide tank mix applied as a pre-plant or pre-emergence treatment to prevent stand loss.
- Application of a seed treatment to prevent insect damage to seeds.
- Use of no treatment to prevent pest losses.

If pest populations threaten to cause stand loss, using a soil insecticide or seed-applied insecticide may be warranted. If corn rootworms threaten to attack corn root systems, using a soil insecticide, high rates of seed-applied insecticide, or Bt hybrid for rootworm at planting may be warranted. Although the use of soil insecticides at corn planting time represents the leading use of insecticides, the Midwest, and North America, 75per cent to 80per cent of the corn is planted without the use of a soil insecticide because of limited pest pressure and extensive annual crop rotation. A review of the conditions that warrant the use of a soil insecticide is presented here.

Pest activity that may cause stand losses and can be prevented by appropriate selection of soil or seed-applied insecticide treatments includes seedcorn maggots, cutworms, grubs, and wireworms. Each of these pest problems is often associated with unique conditions and should not be considered a widespread problem applicable to all corn-growing habitats.

Seedcorn maggots may be a problem when high levels of organic matter are associated with climatic conditions delaying seed germination and seedling emergence. Most problems will occur when organic matter, such as an old alfalfa field or a cover crop, is incorporated into the soil in early spring. Either applying a soil insecticide or a seed treatment may prevent seedcorn maggot injury. In general, if corn is planted into a high organic habitat that is soil incorporated and where adverse climatic conditions may result in delayed emergence, then preventive treatment should be implemented.

Cutworm problems are generally associated with the black cutworm, which is an annual migrant from the South. Infestations of black cutworm depend on whether significant populations are immigrating from the South, and whether a certain field is an attractive location to the immigrating populations of cutworm. In general, most corn fields under conventional or minimal tillage having minimal weed problems are not attractive to migrating cutworms and do not require preventive treatment. No-tillage corn plantings tend to have a higher incidence of cutworm infestations, but it is questionable whether routine preventive treatment for cutworm in such habitats is needed. Stand loss resulting from cutworms may be prevented by applying soil insecticides having a high level of efficacy against cutworm or by applying insecticide as a tank mix in pre-plant or pre-emergence herbicides. The tank mix treatment may also prevent problems associated with common stalk borer, which most soil insecticide treatments do not prevent and which is difficult to control with rescue treatments.

Wireworm and grub infestations associated with stand loss in corn are relatively uncommon, but such problems should be expected when corn follows pasture, sod, forage, or fallow ground. In these production environments, the establishment of specific grasses may facilitate development of wireworm or grub populations with life cycles extending beyond a year. In such situations, using a soil or seed-applied insecticide having efficacy against wireworms and grubs is recommended. A planter box seed treatment may be regarded as a minimal treatment against potential wireworm damage to seed, but using a soil insecticide at planting provides extended protection of both the seed and the seedlings.

Reducing losses resulting from corn rootworm on continuous corn or first-year corn rootworm fields is the predominant objective of most soil insecticide applications, high rates of seed treatment, or Bt hybrids for rootworm. In general, most growers routinely apply soil insecticide to all continuous corn plantings. Although applying a soil insecticide to continuous corn planting is a routine practice, many continuous corn fields do not have rootworm activity warranting preventive treatment. The potential for rootworm injury can be predicted by monitoring adult rootworm activity in the field programmed for planting of continuous corn or using sticky traps in soybean fields where first-year corn rootworms may be a threat the following year. However, adult rootworm activity is rarely monitored in continuous corn. Routinely applying soil insecticides, high rates of seed treatments, or Bt hybrids for rootworm on continuous corn may be justified, however, because the combination of stand loss prevention plus root injury from rootworm may generate a benefit exceeding the cost of application. However, if either stand loss or rootworm injury is lacking, then the benefit derived from these treatments may not outweigh the cost of treatment.

Planting corn without any treatment has been the norm for most corn producers. In general, corn planted without any insecticide treatment is first-year corn under either conventional or minimal tillage without the threat of damage from first-year corn rootworm. Planting no-till, first-year corn acreage, or continuous corn acreage without a soil insecticide, increases the risk of injury resulting from soil pests. Selection of the optimal pest management practice at planting time depends, in part, on a grower's preference for minimizing risk and on personal experience related to losses attributed to pest problems.

Early Stand Loss Prevention

Depending on the degree of preventive treatments taken at planting, corn stands may be vulnerable to infestations that reduce stand establishment. If the early stand loss is a result of seedcorn maggot, wireworm, grubs, or early cutworm infestation, timely response with rescue treatments may not be feasible. However, if a successful stand has been established and significant injury resulting from cutworm, stalk borer, slugs, or armyworm is detected, then application of a timely rescue treatment may prevent further stand losses. Timely detection and response to early pest infestations of corn depends on implementation of a programme of field inspection to detect pest problems in their early stages of development. In general, as tillage decreases, the potential for early pest problems increases. Significant armyworm and slug infestations are primarily no-tillage corn production problems.

Corn planted no-till into either old hay stands or cover crops of rye is extremely vulnerable to armyworm infestations. Under these conditions, armyworm may cause significant defoliation. Such infestations, however, may be easily controlled with a rescue treatment if the infestation is detected in a timely manner. Any corn planted no-till into a grassy habitat should be monitored weekly and semi-weekly if armyworm activity is detected. Rescue treatment is warranted whenever the level of defoliations appears to be resulting in a loss of stand.

Slug problems tend to increase if acreage remains under no-tillage over an extended period. If defoliation is significant, rescue treatment with approved baits is a viable option. However, if slugs continue to be a problem on a yearly basis, implementation of minimum tillage will tend to reduce the overall problem.

Mid- (V8) and Late- (V12) Whorl Pest Problems

As corn stand is established, the relative incidence of severe pest problems tends to decline. However, periodic inspections should be conducted to detect pest problems. The primary pest problem of mid-whorl corn is infestation by first-brood European corn borer. Less than 5per cent of corn stands exhibit significant infestations of corn borer, and few fields ever have infestations warranting corrective action. However, if an infestation is found where one or more larvae can be prevented from boring into the stalks, then rescue treatment may be warranted. If the farm or field has a history of significant European corn borer problems, then the use of a Bt hybrid with activity against corn borer might be warranted. If planting is made in late May or early June, research has shown a yield increase when a Bt hybrid for corn borer is planted at that time.

In general, the development of a borer in a stalk leads to a 5per cent reduction in yield of the infested plant. Therefore, if whorl infestations and larval presence are detected at significant levels, timely application of a rescue treatment may be warranted.

No relationship has been found to exist between corn borer incidence and tillage. Although no-tillage corn production allows corn stubble with the overwintering larvae to remain in the field, corn borer activity in no-tillage corn is equivalent to that of conventional and minimal tillage corn.

Another insect problem, which becomes evident during the mid- to late-whorl stage of corn development, is root lodging resulting from corn rootworm. Although the problem cannot be corrected once it occurs, July is the optimal time to evaluate rootworm impact on continuous corn. If rootworm activity is significant, root-lodged corn plants will be evident. Root systems should then be inspected to confirm and evaluate the degree of rootworm activity.

Tassel and Silk Stage Pest Activity

As corn enters the tassel and silk stages, a few pest problems may occur that may possibly disrupt successful pollination and adversely influence yield. These pest problems include the corn leaf aphid and various beetles feeding on the silks.

Corn leaf aphid problems are generally associated with dry conditions. If abundant, colonies of corn leaf aphids may secrete sufficient honeydew to disrupt the release of pollen from the tassels. Abundant corn leaf aphids on 70per cent or more of the stand may warrant rescue treatment, but the activity of beneficial predators usually takes care of such infestations when they occur.

Successful pollination may be affected by excessive clipping of silks by abundant populations of corn rootworm beetles and Japanese beetles. Application of rescue treatments may be warranted when beetle populations are clipping silks to less than a half-inch or when five or more beetles are present per plant.

Pre-Harvest Corn Pest Activity

As corn ears mature, two pest problems may appear that may influence ear formation and warrant attention. The most common late-season pest problem is infestation of the ears and stalk by the second brood of the European corn borer. Early detection of significant second brood corn borer infestations is difficult except when preceded by abnormal levels of adult moth flight activity.

If significant infestations of second brood corn borer are detected early, rescue treatment may be warranted to prevent excessive stalk injury and ear infestation. When severe infestations of second brood corn borer activity are detected in the stalks, fields should be flagged for early harvest; significant lodging may occur if harvest is delayed. If the farm or field has a history of significant European corn borer problems, then the use of a Bt hybrid with activity against corn borer might be warranted. If planting is made in late May or early June, research has shown a yield increase when a Bt hybrid for corn borer is planted at that time.

Another pest commonly observed in the ears during the late summer is the corn earworm. Although this pest is regarded as an important pest of sweet corn, it is not generally regarded as a significant pest of field corn warranting rescue treatment.

Disease Control

Major corn diseases include leaf blights, stalk rots, ear rots, and kernel rots. Although some diseases can be controlled by a single practice, such as planting a resistant hybrid, most diseases require a combination of practices to ensure that economic damage is kept to a minimum. Once a disease has been identified, its management depends on understanding its cause(s), the factors that favor disease development, which plant parts are affected, and when the disease organisms are spread.

1. Plant high-quality seed, treated with a fungicide seed treatment, in a well-prepared seedbed. Plant seed 1.5 to two inches deep at rates recommended by the seed company to ensure proper plant populations. When populations are excessively high, the stress caused by plant-to-plant competition may increase stalk rot and lodging.
2. Plant high-yielding hybrids with resistance to leaf blight, ear rot, and stalk rot diseases. Several potentially destructive diseases now cause only minor losses because of the widespread use of resistant hybrids. Review the level of resistance available in hybrids offered by your seed dealer before ordering seed for planting.
3. alanced fertility is the key to vigourous, well-developed plants. High rates of nitrogen, especially when excessive in relation to potassium, favor the development of stalk rot and some leaf diseases. Use recommended levels of N, P, and K based on soil tests.
4. Crop rotation and destroying corn residues by tillage reduce the numbers of disease organisms surviving in the field. However, reduced tillage should be practiced to conserve energy and to protect soil from loss through erosion. When corn is planted after corn, especially under reduced tillage production, these disease-management practices are lost. Other disease-management practices, such as growing highly resistant hybrids, become essential to compensate for this loss.
5. Improve soil drainage in poorly drained soils. This reduces water stress and reduces losses from seedling

blights, root and stalk rots.

6. Control insects and weeds in and around fields. Insects such as rootworm and stalk borer create wounds that serve as entry points for fungi causing anthracnose stalk rot. The corn flea beetle serves as the vector of Stewart's leaf blight bacterium. Some weeds act as reservoirs for corn pathogens.
7. Survey fields for damage from leaf blight disease by tasseling. Fungicide application may be justified in commercial corn production fields only if susceptible hybrids are grown. Popcorn and inbreds grown for seed production are generally more susceptible to leaf diseases than dent corn and should be scouted for leaf diseases regularly.
8. Survey fields in the fall prior to harvest to determine the incidence of stalk rot. A rapid and easy technique to determine the incidence of stalk rot is the squeeze method. Grasp the base of the stalk above the brace roots and squeeze the stalk between the thumb and first two fingers. Stalks with significant rot will crush easily. Those fields with the greatest percentage of rotted stalks should be harvested first to avoid losses resulting from lodged corn.
9. Proper adjustment and operation of the combine or picker reduces harvesting losses in the field with stalk-rotted, lodged corn. Some equipment companies have attachments for the combine header to help pick up lodged corn.
10. For long-term storage, dry shelled corn to 13per cent to 14per cent. Ear corn to be cribbed should be dried to 20per cent moisture. Maintain cool and dry storage conditions to prevent storage molds from developing.

HARVESTING

Harvest date should be determined by crop maturity, not by the calendar. Plan to harvest fields with potential lodging or harvest loss problems (such as stalk rot or deer damage) first. All field shelled corn with more than 15per cent moisture must be dried for safe storage. The ideal kernel moisture level at which to harvest for dry grain storage is 25per cent. Corn normally dries approximately 0.75per cent to 1per cent per day during favorable drying weather (sunny and breezy) during the early, warmer part of the harvest season —from mid-September through mid-October in central. By late October to early to mid-November, field dry-down rates usually drop to probably no more than 0.5per cent per day. By mid- to late November the rate drops to 0.25per cent per day, and after Thanksgiving, drying rates are negligible.

Dry-down rates can also be estimated in terms of Growing Degree Days (GDDs). Generally it takes 20 to 30 GDDs to lower grain moisture each point from 30per cent down to 20per cent. In September, accumulation of GDDs averages 10 to 15 per day. In October, the accumulation drops to five to 10 GDDs per day. These estimates are based on generalizations, however, and some hybrids may vary considerably from this pattern of dry-down.

Monitoring harvest losses is an important part of the harvesting process. Ear corn losses from in front of the combine (preharvest losses) should be subtracted from the total harvest loss estimate. The loss of one normal-sized ear per 100 feet of row translates into a loss of more than one bushel per acre. An average harvest loss of two kernels per square foot is about one bushel per acre. Keep in mind that most harvest losses occur at the gathering unit.

Drought-induced stalk lodging and insect problems reduce the yield potential of many corn fields if harvesting is delayed much beyond maturity. Ear drop damage may be high in some years as a result of extensive European corn borer damage. Estimates of harvest losses based on long-term average data at Purdue University indicate that losses increase by 1per cent to 2per cent for each week of harvest delay. Ear damage by corn borers and other insects may also increase the potential for grain quality problems caused by ear molds.

Shelled grain weights can be adjusted using a grain shrink table. Shrink represents both the moisture loss and a 0.5per cent dry-matter loss encountered during dry and grain handling. To estimate the amount that a given wet weight of corn will lose during the drying process, multiply the wet weight by the shrink factor from the table. For example, assume that one ton (2,000 lb) of shelled grain at 25per cent moisture will be dried to 15.5per cent moisture. Drying and handling losses are 2,000 lb × 0.1174, or 235 lb. This results in 2,000 lb minus 235 lb, or 1,765 lb of grain at 15.5per cent moisture. Monitor debris and cracked corn in the grain as harvesting progresses. Debris and cracked corn lower grain quality and increase the potential for spoilage of stored corn.

Test Weight

Test weight of corn determines the weight of a bushel volume of grain. Test weights determined on dry corn indicate whether the grain crop reached full maturity. Low test weights indicate immaturity. If bushel test weight of mature corn is determined at harvest when grain moistures are greater than 15.5per cent, the test weights will be biased downward. In other words, as corn grain dries, test weight increases. Differences in test weight influence USDA grading of shelled corn. The adjustments in test weights do not apply if grain contains more than 10per cent broken kernels, was damaged by drought or disease, was harvested when immature, or was dried at air temperatures of 180°F or higher.

Ear Corn

Ear corn can be cribbed safely when the grain moisture is 21per cent or less. However, with cold weather and narrow (four-foot), well-ventilated cribs, corn may be stored when grain moisture is several percentage points higher. Use Table to convert ear corn yields to shelled corn equivalents. For example, four tons (8,000 lb) of ear corn at 21per cent grain moisture is equivalent to 8,000 divided by 77.6 or 103 bushels of shelled corn.

Corn Silage

Corn harvested for silage yields one-third more feed nutrients per acre than corn harvested for grain. Corn in the full dent stage produces 50per cent more feed than in the milk stage and 100per cent more feed than in the silking stage. Corn harvested in the milk or silking stage results in poorer quality silage because of its high moisture content.

One of the most important steps in producing quality corn silage is to harvest at the proper moisture. The storage structure determines the proper moisture level at which to harvest.

Desired moisture levels for different structures are as follows:

Sealed airtight silos—55per cent to 60per cent

Bag silos—60per cent to 70per cent

Upright silos—62per cent to 68per cent

Trench silos—65per cent to 70per cent.

Ideally, moisture levels in the silage should be monitored at harvest to prevent harvesting the crop outside the desired range. If moisture testing is not feasible, then estimate the crop moisture by the stage of crop development.

Kernel milk line can serve as an indicator of whole plant moisture levels. As kernels start to dent, a separation between kernel starch and milk can be seen. The firm starch is deposited in the crown (outer) area of the kernel, and the milk occupies the basal area of the kernel. This appears as a whitish line separating the two areas. As the crop matures, this kernel milk line moves down the kernel, and the whole plant moisture declines. When this line reaches the midpoint of the kernel, 90per cent of the final kernel dry weight has been achieved, and silage yields reach a maximum. At this point, the stover part of the plant has good digestibility, and the moisture is usually in the desired range for storage in airtight silos.

A higher moisture level and a slightly earlier harvest are recommended for bunker silos and upright conventional silos. Harvest time can be predicted by monitoring the progression of the milk line. When the milk line reaches the base of the kernel, a black layer forms and the crop is physiologically mature.

Silage harvest should not be delayed beyond the black layer point because the silage gets too dry, the kernels tend to harden, and the digestibility of the stover declines rapidly. The desired chopping length for corn silage is 5/8 to ¾ inch. If silage is harvested when the crop moisture is lower than desired, consider chopping finer than normal to promote good packing and to minimize air pockets in the silage.

Specialty Types of Corn

High grain and silage yield potential, high feed value, and availability of adapted superior hybrids account for the widespread use of yellow dents. Yellow dents have the highest content of carotene (vitamin A) of the cereal grains. Other types of corn include flint, pop, waxy, and sweet. Each type of corn has unique kernel characteristics that determine its use and how it is grown.

Because most specialty corn hybrids (including white dent, waxy, high oil, and popcorn) are grown under contract, it is advisable to identify a market before planting. Also, some specialty corn processors specify certain hybrids and cultural practices they want growers to use. Contracts for growing specialty hybrids usually offer a premium over the yellow dent price to compensate for the lower yield potential and the special handling required to ensure high grain quality. Most specialty corns must be grown in isolation from conventional corn fields to prevent cross pollination.

White Corn

White corn types are equal to yellow types in carbohydrate content but are deficient in vitamin A. White types are grown primarily for direct human consumption in foods such as breakfast cereals and grits. Yields of white hybrids generally are not competitive with yellow dent yields. Typical high-yielding yellow dent hybrids produce 5per cent to 10per cent more grain yield than the best white hybrids. White corns must be grown in isolation from yellow dents to prevent cross-pollination. A mixture of white and yellow downgrades either type at the market.

Waxy Corn

The carbohydrate or starch granules of regular dent corn consist of approximately 75per cent amylopectin and 25per cent amylose. Waxy corn has nearly 100per cent amylopectin. Waxy corn was initially recognized as a valuable source of industrial starch. The stability and clarity of amylopectin starch make it highly suitable as a food thickener. Waxy corn has also been considered as a potential animal feed. Feeding trials with waxy corn hybrids have occasionally shown a benefit, but they have not been consistent. Changes in feed efficiency and production have been insignificant, but usually in favor of the waxy corn hybrids. Yields of waxy corns are generally lower than those of yellow dent corn.

Growers should have some estimate of the relative yield performance of specific waxy hybrids before growing any substantial acreage. Waxy corns must be grown in isolation from other corn types to maintain purity standards for industrial use. Keeping the harvested grain separated by 12 to 16 border rows from the rest of the field is considered adequate isolation; 5per cent contamination with regular corn pollen is considered the upper limit.

High-Lysine Corn

Although normal dent corn contains about 9.4per cent protein, the quantity of two essential amino acids, lysine and tryptophane, is below nutritional requirements for humans and nonruminant (single-stomached) animals, such as pigs and chickens. High-lysine corn corrects this deficiency and may be advantageous in swine feeding rations. Adapted varieties of high-lysine corn are limited in number and have generally been lower in grain yield than normal dent varieties. The softer kernels of high-lysine corn are more vulnerable to breakage at harvest, which can lead to higher incidence of kernel or ear rot. Production fields must be grown in isolation from other corn fields to ensure protein quality. Cross pollination results in normal dent corn. A separation distance of 300 feet is recommended.

High-Oil Corn

High-oil corn (HOC) is attractive as a livestock feed because it has greater energy value than normal yellow dent corn and can replace more expensive dietary sources of fats and proteins. HOC contains 50per cent to 100per cent more oil than normal yellow dent corn, which averages about 4per cent oil on a dry weight basis. Feeding trials with HOC indicate that it has improved feed efficiency and results in increased rate of gain over conventional corn.

The TopCross grain production system, currently the most widely used method of producing HOC in the United States, involves the planting of a blend of two types of corn. One type, referred to as the grain parent, is a cytoplasmic male sterile (produces no viable pollen) elite hybrid that comprises 90per cent to 92per cent of the seed in the blend. The second type, comprising 8per cent to 10per cent of the seed in the blend, is a high oil pollinator. The high oil pollen contains genes that cause production of kernels with larger than average embryos. Because most of the oil and several essential amino acids are in the embryo, the increased embryo size of HOC results in greater oil content and enhances grain protein quality.

Following the introduction of the TopCross system, HOC production in the United States increased from less than 50,000 acres to more than one million acres. However, since then, HOC production has dropped sharply with acres planted in 2002 estimated at less than 500,000. A major factor contributing to this decline has been the availability of cheaper sources of oils and fats which compete with HOC as an energy source in livestock feed rations. In TopCross HOC production, the minimum isolation distances recommended by seed companies range from 60 to 200 feet.

Popcorn

Popcorns are essentially small-kerneled flint corns and are among the most primitive of the surviving races of corn. Kernels contain a hard endosperm with only a small portion of soft starch. Popcorns are generally either pearl or rice types. Pearls have smooth, rounded crowns, and rices are pointed. Heating the kernel turns the moisture inside the soft starch in the center into steam that explodes the kernel inside out. The greater the expansion, the higher the quality. Hybrids differ as to kernel quality, which also includes flavour, tenderness, absence of hulls, colour, and shape. Popcorn is one of America's most popular snack foods. American consumers eat, on average, 17 billion quarts of the healthy treat each year. Popcorn is not only high in fibre, potassium, and vitamins, it is also low in calories and fat. A cup of unsweetened popcorn contains between 31 and 55 calories.

Popcorn hybrids usually yield less than half of normal dent hybrids. To achieve maximum quality, minimize mechanical damage and dry with low heat to a moisture of 13.5per cent. Overdrying and kernel damage result in reduced popping volume. Handling and quality are extremely important aspects of popcorn production. From an agronomic standpoint, popcorn must be planted to mature before frost, and herbicide programmes must be labeled for popcorn. Use a fertility programme for a 150 bushel per acre yield goal and pay extremely close attention to potassium fertility to guard against poor stalk quality and lodging. Popcorn production fields do not have to be isolated from other types of corn, even if the popcorn is not dent-sterile. However white and yellow popcorn should not be planted within 0.5 mile of each other to minimize cross-pollination.

Isolation Requirements for Identity Preserved (IP) Non-GMO Corn Production

Managing pollen drift is an important consideration in the production of specialty corns and non-GMO corn as IP grain crops. Corn is a cross-pollinating crop in which most pollination results from pollen dispersed by wind and gravity. Although most of a corn field's pollen is deposited within a short distance of the field, with a 15-mph wind pollen may travel as far as one-half mile in a couple of minutes. Pollen from corn containing transgenes or genetically modified organisms (GMOs), such as Bt corn, may contaminate (by cross-pollination) nearby non-GMO corn.

The European Union guidelines require that foods, including grains, containing more than 0.9per cent biotech material (GMOs) be labeled as genetically engineered. Producers of IP non-GMO corn, such as organic farmers, need to minimize pollen contamination by GMO corn if they are to obtain premiums. Growers can follow several planting practices to minimize GMO pollen contamination, including use of isolation and border rows, planting dates, and/or hybrid maturity.

Chapter 6

Soybean Production

The major objective of a crop production system is the interception, fixation, and storage of sunlight energy. There are many components of a system that will accomplish that objective. The most important are early planting; narrow rows; productive varieties that resist disease; control of weeds, insects, and diseases that rob energy from the system; and availability of soil nutrients in adequate amounts. Other inputs to the system must not limit energy fixation or slow the process. The effects, interactions, and relationships of various inputs of an efficient soybean production system are discussed here.

VARIETY SELECTION

Most soybean varieties have genetic yield potentials well over 100 bushels per acre. A variety's adaptability to the environment and production system where it will be used sets the yield potential of the production system. The quality of the weather during the growing season and the stresses from weeds, diseases, and insects determine what the crop yield will be. A variety's performance in a previously conducted yield trial is a measure of its performance in that particular environment and production system and does not assure satisfactory performance under a different set of conditions.

When a group of varieties is tested for yield over a range of environments, their rank order commonly changes, which indicates that some varieties are better adapted to a specific environment than others. Therefore, it is best to select varieties with characteristics that will help them perform well in the cultural system and environment to be used, rather than on their yield record alone. For example, if excessive growth and lodging are problems, then select varieties that are medium to short in height with good standability. If the field has a history of *Phytophthora*, then select a variety with a resistance gene plus a high partial resistance rating to address that problem. The selection of medium or small seed when using a grain drill will improve metering and stand uniformity.

Alternatively, one can select the varieties that performed best at a test site that is similar to the field for which a variety is being selected, or select a variety that has performed well over several test sites that vary widely in yield potential. Maturity information should be used to select varieties that mature at different times to allow for timely harvest and high test weights. Generally, each 10-day delay of planting in May delays maturity three to four days in the fall. For best yields in wide rows, select full-season varieties with a bushy growth habit. Growth habit is not important in narrow rows. Fitting the variety to the environment is superior to selecting a variety and hoping the environment and weather will fit it.

Variety Performance Trials

The purpose of the Soybean Performance Trials is to provide an unbiased evaluation of variety characteristics and performance to facilitate the selection of varieties appropriate for particular production sites and systems. Field trials are conducted at six locations representing the diverse production.

Early Season and Seed-Borne Diseases

Phytophthora root and stem rot is the most serious soybean disease and is present everywhere soybeans are grown. Damage to the crop by *Phytophthora* is most prevalent in fields with poor drainage, high number of years with soybeans, and reduced tillage systems. Varieties are susceptible at all stages of growth. Saturated soil with a temperature above 60°F provides the ideal conditions for infection. Susceptible varieties should not be grown in poorly drained soils or on soils known to have a history of the disease. Seed of varieties with good partial resistance should be treated with a fungicide that aids in the control of *Phytophthora* damping off. Varieties with *Rps* resistance gene(s) should also be treated to control *Phytophthora* damping off because these *Rps* genes are not effective in every field nor across the whole field. Planting early, well before soil temperatures reach 60°F, often allows varieties with high levels of partial resistance to escape early infection if the soil does not become saturated.

Fig. Soybean Variety Performance Trial plots.

Pythium and *Rhizoctonia* root rots are also common, and most varieties are susceptible. Damage to plant stands is greatest on poorly drained soils and during seasons of high rainfall. *Pythium* is controlled by Apron XL and Allegiance FL seed treatments, and seedling infections of *Rhizoctonia* are controlled by a number of seed treatment materials.

Phomopsis seed rot can be severe when rainfall occurs intermittently during grain drydown and harvest. The longer soybeans are in the field after ripening, the greater the incidence of seed rot. Harvesting soon after the soybeans mature decreases the amount of seed damage. Using varieties with a range of maturities allows for a more timely harvest of each field. Many varieties are resistant to *Phomopsis* seed rot; if *Phomopsis* develops in a variety, look for a different variety for future years. Fungicides that are currently labeled for soybeans have little activity for *Phomopsis* and are generally not economical. Crop rotation and tillage are excellent management tools for this seed-rot pathogen as it survives on old crop residue.

Phomopsis seed rot can reduce overall germination in certain seed lots and *Phytophthora, Pythium,* and *Rhizoctonia* can kill seeds and seedlings after they are planted. One of the management tools for these seed and soil-borne plant pathogens is to use fungicide seed treatments. The efficacy of many of these compounds. Note, no one fungicide is highly effective for all pathogens. Choosing a mix of several compounds will provide broad spectrum control. Seed treatments are best used on fields with poor drainage, a history of stand establishment problems, and reduced tillage systems.

Mid-Season to Late-Season Soybean Diseases

Soybean cyst nematode (SCN) now exists in production fields throughout. In some fields, the population of SCN is currently quite high. Populations of SCN may take eight to 10 years from introduction to reach damaging levels throughout a field. In a variety test in west central on a fertile, dark-coloured soil, varieties resistant to SCN yielded more than 50 bushels per acre, whereas those susceptible to SCN yielded from 24 to 39 bushels per acre. Although these studies were conducted in problem fields, the estimated yield loss from SCN in other Midwestern states is 8per cent to 12per cent.

In the vast majority of fields, SCN causes no above-ground symptoms. The only difference that growers will see is that yields may be five to 10 bushels less than fields with similar yield potential. In more severe situations, where SCN populations are high, injury is easily confused with other crop production problems, such as nutrient deficiencies, injury from herbicides, soil compaction, or other diseases. The first field symptoms are usually detected in circular to oval patches of stunted, yellowed plants. Symptoms are most evident in late July or August when plants are under drought stress or in fields with low fertility. When populations of nematodes are high, the symptoms may even occur under normal to optimal growing conditions. Affected areas of a field may increase in size each year in the direction of tillage. In these affected areas, SCN females can often be found feeding on the roots.

SCN is best managed with crop rotation, rotating non-host crops such as wheat, corn, alfalfa, or red clover and rotating sources of SCN resistance. Never plant a SCN-resistant variety without checking your SCN population levels first. When a non-host crop is planted, SCN populations will decline by as much as 50per cent annually. SCN resistance is measured by a reduction in the number of females that feed on roots, but a few females will reproduce. Thus, over time, populations will adapt to these sources of resistance and reproduce in increasing numbers. Sources of resistance that are currently available to soybean producers include PI88788, PI437654 (Hartwig came from this source), and Peking.

To determine what your SCN levels are, soil samples should be taken from the top four inches of soil. Each field should be divided into sections not exceeding 10 acres, and each section should be sampled by taking 15 to 20 subsamples in a zigzag pattern. This level of sampling is necessary to obtain relatively accurate counts of the nematode population and to make meaningful recommendations for control. The soil samples should be moist, but not wet; packaged in double plastic bags; and protected from becoming too warm. Mail samples to the C. Wayne Ellett Plant and Pest Diagnostic Clinic,

Phytophthora stem rot will continue to infect plants throughout the growing season. This late-season phase of the disease can only be found in fields where heavy rains or saturated soils have occurred and with varieties having ineffective *Rps* genes and low levels of partial resistance. If the *Rps* genes are effective against the P. sojae population, then no disease will develop; however, if they are no longer effective, then stem rot will develop. We have found from a number of years and locations that varieties with high levels of partial resistance rarely develop stem rot. One reminder is that not all seed companies use the same scoring system.

Sclerotinia stem rot is present throughout and may be severe (50per cent of plants in a field infected) when wet weather occurs prior to and during flowering. Varieties with resistance to *Sclerotinia* have fewer numbers of plants infected but all are susceptible to some degree. Stem symptoms first appear as water-soaked lesions followed by cottony growth and, eventually, black irregular-shaped sclerotia which resemble mouse droppings in form. Wide rows (30 inches) aid in control by permitting air to move through the canopy to dry plant leaves and the soil surface but also reduce yield due to less sunlight fixation. Reduction in plant populations (160,000 to 180,000) and planting in 15-inch rows can reduce the overall incidence of *Sclerotinia* stem rot without negatively impacting yields.

Brown stem rot can severely reduce yield. This fungus enters the plants through the roots and slowly colonizes the stem and the xylem, where it interferes with water transport. The disease symptoms develop after flowering and are identified by an internal browning of the stem in August. Foliar symptoms are rarely seen, but the leaves of infected plants may suddenly wilt and dry 20 to 30 days before maturity and drop from the plant. Crop rotation is an excellent control for this disease.

Sudden Death Syndrome (SDS) is another late-season disease that appears to always be associated with soybean cyst nematode and areas of the field with very poor drainage. Symptoms are very similar to brown stem rot in that brown spots develop in the leaves between the veins, and this is surrounded by a bright yellow chlorosis. In SDS, the roots are very degraded along with the crown. One of the key diagnostic tools is the colour of the pith, which remains white and healthy with SDS and is brown and decayed with brown stem rot. This fungus survives in the soil for long periods of time, so to prevent rapid build-up of the pathogen, crop rotation and improving soil drainage are key.

Crop Rotation

Crop rotation is the most effective pest-control practice available to crop producers. The sequence of crops grown in a field affects the productivity of each crop. Research from most Midwest states indicate that a soybean crop following a crop other than soybeans will usually yield about 10per cent more grain, on average, than when soybeans follow soybeans. Many of the crop disease and insect problems currently experienced are due to short crop rotations or no crop rotation. If all our crops were produced in a four-year crop rotation, yield loss to disease and insects would be near zero rather than at the 8per cent to 12per cent we currently experience.

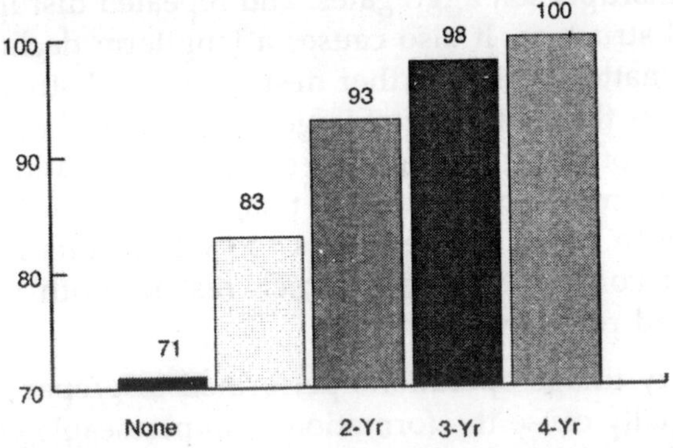

Fig. Effect of length of crop rotation on percent crop yield.

The effect of crop rotation on yield has been thoroughly investigated by most Land-Grant Universities. Researchers at the University of Wisconsin conducted a crop rotation study with corn and soybeans. Averaged across 29 years by location environments, corn and soybeans in their corn-soybean rotation resulted in 13per cent and 11per cent greater yield than the respective monoculture. Kansas State University conducted a 20-year crop rotation study in which the soybean yields were 20per cent greater when rotated with wheat or grain sorghum than without rotation. Results of a Canadian crop rotation study showed that soybean yields from a wheat-corn-corn-soybean rotation were 7.1per cent higher than in a corn-corn-soybean rotation. Results of a 10-year crop rotation study conducted in northern region indicated that continuous corn yielded only 89per cent as much as corn in a corn-oats-hay rotation, and corn in a corn-soybean rotation yielded 94per cent of the corn-oat-hay rotation.

Economics often dictates crop sequence, but where choices are available, soybeans should follow crops other than soybeans, sod, or legume crops like alfalfa. Corn or other grass crops can make good use of the nitrogen left by legume crops and sod. The effect of the length of a crop rotation on yield can be seen in Figure 5-3.

Tillage

Tillage disrupts soil aggregates, and repeated disruption destroys soil structure. It also causes a long-term decline in soil organic matter, which further destabilizes soil structure. Tillage disrupts the continuity of large soil pores and restricts the movement of water through the soil profile, creating soil drainage problems. Repeated use of tillage tools operating at the same depth or when the soil is too wet results in the formation of compacted zones which restrict both water movement and root development.

Secondary tillage operations performed to prepare fine seedbeds usually cause the formation of impermeable crusts on light-coloured silt loam soils such as Blount and Crosby. These crusts reduce seedling emergence, air exchange, and water intake, all of which reduce yields. When thick crusts form, disrupting them by rotary hoeing or cultivation often improves yield, particularly in dry years.

Tillage is one of the largest out-of-pocket expenses used for crop production and often does not generate enough yield to make the tillage profitable. While no-tillage can reduce production costs and increase profits, it also creates problems that producers must solve with proper management of the other inputs and production practices. Some of these problems are colder, wetter soil at planting, more root rot disease, slower emergence and growth, dealing with crop residues and the diseases they contain, and more. There are times when tillage is warranted and will likely be profitable. Here is a partial listing of some of the situations when tillage may be needed:

- Use tillage when inadequate soil drainage leads to serious yield loss due to root rot diseases, poor stand establishment, or late planting.
- Use tillage to bury crop residue and thus reduce pathogen and insect survival that can infect a following crop.
- Use tillage as a prelude to land leveling, rock removal, and for the incorporation of soil amendments, such as lime or very high rates of fertilizer.
- Use tillage to mitigate compacted soil layers or zones that interfere with water movement into and through the soil, which may delay planting, harvesting, and other field operations.

The cost to perform various tillage operations and the yield increases needed to pay for those operations.

Producing Soybeans Without Tillage

Growing soybeans without tillage has become both practical and profitable and often reduces or eliminates some tillagerelated problems. Time savings accrued by eliminating tillage can be invested in earlier and more careful planting or the planting of more acres. Maintenance of crop residue on the soil surface reduces soil crusting, which can lead to better and more uniform seedling emergence; improved yields on some soils; and reduced needs for rotaryhoeing, cultivating, and replanting. In addition, notillage systems do not bury weed seeds, reducing the germination potentials of some species, particularly largeseeded broadleaf weeds. Finally, use of notillage systems can prolong the life of surface drainage improvements, particularly on flatter fields.

When planting notillage soybeans, growers should pay attention to soil drainage; planting procedures; crop rotation options; and disease, insect, and weed control. While the improper management of any of these factors will reduce yields in tillage systems, their effects can be much more adverse with no tillage. Problems created by removing tillage from the crop production system and important adjustments that must be made to offset those negative effects are presented here:

Cooler Soil Temperatures Slow Germination, Emergence, and Early Growth

Because the soil is warmer at the surface than at the 1.5-inch planting depth, the solution is to plant shallow (one inch), but in moist soil. The warmer soil temperatures at shallow depth will enable seeds to germinate and emerge earlier and, in effect, produce a closed leaf canopy and get to the reproductive stage sooner. The use of narrow rows (7.5 inches) will compensate for the slower early growth associated with no-till production. Good seed-to-soil contact and the use of high-quality seed treated with the appropriate fungicides promote the rapid emergence of a healthy crop. Slower planting speed will allow the planting tool to space seed more uniformly in the row and at a more uniform depth, so that seeding rates can be reduced and production costs can be lowered.

Root Rot Diseases are much more severe due to a Wetter and Cooler Soil Environment

Two actions can increase plant stands and improve root health:

Select varieties with high levels of *Phytophthora* partial resistance that will give a good level of protection against all strains of *Phytophthora*.

Treat the seed with Apron XL or Allegiance to protect the seedling from *Phytophthora* and *Pythium* root rot until the partial resistance mechanism takes effect just after emergence.

Another strategy is to use soybean varieties that have one or more *Rps* resistance gene(s) for control of *Phytophthora* root rot. Other broad spectrum fungicides will control other diseases that damage the root system and lower stand counts.

Do not use no-till culture on poorly drained fields, and do not plant when the soil is too wet for shallow tillage. Tillage or planting operations on a wet soil compact the soil particles, which inhibits the proper development of root systems and thus reduces yield

Heavy crop residue and more dense soil can interfere with proper seeder function and lead to poor distribution, poor placement of seed, and lack of adequate depth control.

Spreading crop residue evenly when harvesting will help keep the field surface uniformly covered with residue and at uniform moisture so the entire field is ready to plant at the same time. Due to its fineness, wheat residue keeps the soil colder and wetter than other residues because it provides nearly 100per cent cover and its light colour reflects sunlight. Remove wheat straw when possible and partially incorporate the stubble with a disk to promote its degradation.

Try not to plant on old corn rows since old corn roots interfere with depth control and seed placement. Maintain a down pressure of at least 200 pounds on each row opener to penetrate hard soil areas and use a good depth-control mechanism to maintain the proper seeding depth in soft soil. A residue cutting colter will prevent hair pinning of residue into the seed furrow, which interferes with seed placement, and it will also loosen some soil that the furrow closers can use to cover seed and improve seed-to-soil contact.

Rhizobium Inoculation

Fifty-eight soybean inoculation trials were conducted between 1995 and 2004. During that time, the average yield increase due to inoculating was about 2.0 bushels per acre. The cost of inoculating an acre of soybeans is $2 to $3, depending on the product and the rate used. If soybeans are worth $7 per bushel, the per-acre profit for inoculating soybeans would be about $11 to $12 per acre or a return on investment of more than 300per cent.

Dry and liquid formulations of the same product appear to perform similarly. Once the carrier of the inoculum dries on the seed, the bacterial cells start dying. Inoculated seed should be planted as soon as possible after treatment (12 hours or less), so the bacterial cells will remain moist and survive long enough to infect soybean roots following germination.

When applying a fungicide or using fungicide-treated seed, be sure the fungicide has dried before applying inoculum to the seed. Currently, most inoculation products may NOT be mixed with fungicides and applied to seed together. One exception is that liquid formulations of inoculum may be mixed with ApronMaxx RTA fungicide and applied together. Work is underway to develop formulations of additional fungicides that can be mixed and applied with inoculation materials. Development of inoculation materials that can be applied up to several weeks before planting is also underway.

When loading a drill or a planter using an auger, liquid or dry inoculation materials should be added to the seed as it enters the auger for thorough application. When loading a planter or drill from bags, fill the seed box to a depth of three inches and scatter an appropriate amount of inoculum over the seed and mix thoroughly. Continue to add seed in six-inch layers, treating each until the box is filled. With some dry materials, it may be necessary to slightly moisten seed to increase adherence. A few small specks of inoculum on each seed is adequate. At the recommended use rate, there will be more than 500,000 bacterial cells on each seed. Excessive amounts of inoculum on seed can reduce seed metering by up to 35per cent. Seeding equipment should be calibrated using the treated seed to be planted. Some seeding-rate monitors allow a continuous check of seeding rates, so adjustments can be made to the seeding rate if and when necessary.

When soybeans are planted in a field for the first time, it is not uncommon for even the most ideal inoculation procedures to be less than adequate for producing enough nitrogen for a good crop. When the nodules are insufficient to supply adequate nitrogen, it will be necessary to supply some nitrogen to the crop. In this event, one application of 75 pounds actual nitrogen as urea can increase yields by 8 to 12 bushels per acre. This supplemental nitrogen should not be applied until flowering, which is usually late June and July depending on variety maturity, date of planting, and the weather. To assure the establishment of a reliable inoculation for future years, it is advisable to grow soybeans in a new field for two successive years and to inoculate the seed thoroughly both years.

For satisfactory nitrogen fixation in eastern region where soils tend to be more acid, the pH in the plow layer should be above 6.5, and the percent base saturation of calcium and magnesium should be greater than 40per cent and 10per cent, respectively. On fields where the lime requirement is very high, a shallow incorporation (two to four inches) of two to four tons of dolomitic limestone will aid in the establishment of bacterial colonies on the root system. Dolomitic limestone should be used whenever magnesium levels are lower than 10per cent base saturation.

Planting Date

Regardless of planting date, row width, or plant type, the soybean crop should develop a closed canopy (row middles filled in) prior to flowering or by the end of June, whichever comes first. Generally, when planting in early May, rows must be less than 15 inches apart to form a canopy by late June. An early canopy results in high yields because more sunlight is intercepted and converted into yield than when row middles do not fill in until late in the growing season. Assuming a half bushel per acre per day yield loss with delayed planting, a 10-day delay in planting 300 acres would decrease total production by 1,500 bushels, which is worth approximately $10,500.

Adequate, vigourous stands are sometimes more difficult to obtain with early planting. Seed treatments, good seed-soil contact, and reduced seeding depths, however, aid in establishing vigourous stands. Herbicide programmes must provide weed control for a longer time until the crop is large enough to suppress weed growth through competition. Narrow rows provide the needed competition for weeds sooner than wide rows.

Table: Effect of Row Spacing on the Number of Days (and Date) to Complete Canopy Formation.*

Row Spacing	Date of Planting		
	Before May 5	May 6 to 15	May 16 to 25
7	35 (6/5)	30 (6/10)	25 (6/15)
10	40 (6/10)	35 (6/15)	30 (6/20)
15	50 (6/20)	45 (6/25)	40 (6/30)
20	60 (6/30)	55 (7/5)	50 (7/10)
30	75 (7/15)	70 (7/20)	65 (7/25)

* Assumes very good growth conditions in May, June, and July. Poor growing conditions increase the amount of time needed to complete canopy formation.

Late Planting

Late planting reduces the cultural practice options for row spacing, seeding rate, and variety maturity. The row spacing for June planting should be no greater than 7.5 inches. Appropriate seeding rates for the first half of June are about 200,000 to 225,000 seeds per acre. For the last half of June, 225,000 to 250,000 seeds per acre is recommended, and in early July, the recommendation is 250,000 to 275,000 seeds per acre.

Relative maturity (RM) has little effect on yield for plantings made during the first three weeks of May, but the effect can be large for late plantings. During the first half of June, a four-day delay in planting delays physiological maturity about one day. In the last half of June, it takes a five-day planting delay to delay physiological maturity one day. As planting is delayed, yield potential goes down, and there is concern about whether late maturing varieties will mature before frost.

When planting late, the rule of thumb is to plant the latest-maturing variety that will reach physiological maturity before the first killing frost. The reason for using late-maturing varieties for late planting is to allow the plants to grow vegetatively as long as possible to produce nodes where pods can form before vegetative growth is slowed due to flowering and pod formation. More nodes equates to more pods and more yield. Late-maturing varieties are needed that will mature before getting frosted, but since the first frost date is unknown, we use a narrow range of maturity that will not be damaged by frost occurring at the normal time.

The recommended relative maturity ranges assume normal weather and frost dates, so varieties with those relative maturities should mature before frost and produce maximum possible yields when planted on the dates indicated. Varieties with an earlier relative maturity will mature earlier but will produce reduced yields.

Row Spacing

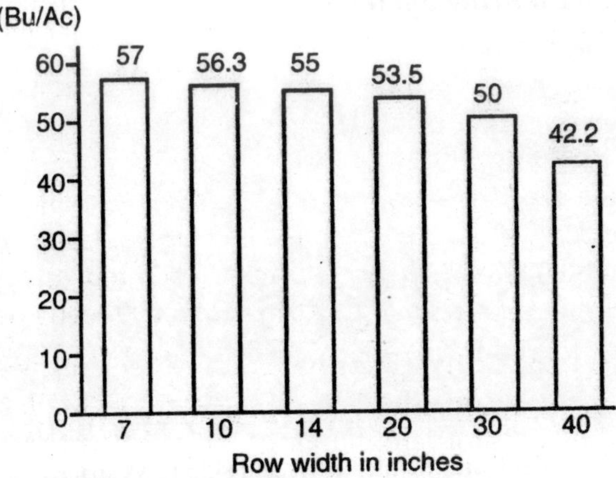

Fig. Effect of row spacing on soybean yield.

Most soybeans are planted in narrow rows. Soybeans grown in narrow rows produce more grain because they capture more sunlight energy, which drives photosynthesis. Within limits, as sunlight interception increases, so does yield.

Researchers have learned that the peak demand for the products of photosynthesis occurs during the reproductive stage. Therefore, the row width should be narrow enough for the soybean canopy to completely cover the interrow space by the time the soybeans begin to flower (June 20 to July 10). The row widths that will accomplish that goal will vary with soil type, planting date, weather conditions, and, in some cases, variety.

The later in the growing season soybeans are planted, the greater the yield increase due to narrow rows. The response to narrow rows is also greater for short varieties and when growing conditions cause plants to grow slowly or be short. The effect of row spacing on the yield for early-planted soybeans can be seen in Figure 5-5. Planting systems using precision seed metering to achieve uniform seed spacing within the row plus uniform depth of seed placement usually produce higher yields than planting systems with less uniform seed spacing and variable depth of planting.

Skip-Row Planting Systems

Fig. A soybean field with 15-inch rows and skip rows spaced 10 feet apart to accommodate application of pesticides.

Modern production systems rely on postemergence herbicide application for weed control and postemergence insecticide application for insect control. This reliance necessitates the use of large equipment for pesticide application in soybean fields during the growing season. Many modern applicators are equipped with narrow tires so that crop damage is minimal early in the season when plants are small, but plant damage increases as plant size increases. Skiprow planting patterns enable the application of pesticides through most of the growing season without causing crop damage. Skiprow systems usually consist of narrow rows with strategically placed wider row spacing to accommodate application equipment tires.

Yield loss due to these wider row middles is usually too small to measure (0.05 bushels per acre). For example: A 60-foot spray boom covers 96 rows spaced 7.5 inches apart. Leaving two of the 96 rows unplanted would reduce the yield by 2.1per cent times the yield difference between 7.5-inch and 15-inch rows. This reduction amounts to, at most, two bushels per acre or 2.5 pounds of grain, which is worth about 30 cents. The yield loss due to running over two rows of soybeans in August is usually $4 to $8 per acre, depending on the yield potential, value of grain, and working width of the sprayer. For example, 2.1per cent times 50 bushels per acre is 1.05 bushels per acre worth $7.35 if we lose all the grain from the two rows run down in August. Additional advantages of skip-row planting systems are the elimination of application skips and overlaps, timely applications, and the adoption of GPS and automated guidance systems. When custom application is planned, consult the applicator for acceptable widths of skip rows, their spacing, and frequency across the field.

Plant Population

When the soybean crop is planted in rows spaced 7.5 inches apart, the effect of plant population on yield is very small over the normal range of seeding rates and for any particular set of conditions. For a crop planted before May 20 in narrow rows, final populations of 100,000 to 120,000 plants per acre are adequate for maximum yield. Final populations for mid-June plantings should be in the range of 130,000 to 150,000 plants per acre. Final populations for early July plantings (double crop) should be greater than 180,000 plants per acre. Final population is a function of seeding rate, quality of the planting operation, and seed germination percentage, and depends on such things as soil moisture conditions, seed-soil contact, disease pressure, fungicide seed treatments, and more. Final harvest stands are typically 60per cent to 80per cent of the seeding rate when high-quality seed is used, and there are few impediments to stand establishment.

Some seed is dead when planted; other seeds may not have the vigour needed to emerge and grow rapidly; some will be lost to disease and insects prior to emergence; some emerged plants will be killed by disease; and some will not grow fast enough to compete adequately for sunlight and thus perish during the growing season. The effect of soybean seeding rate on yield and harvest population for two very different environments can be seen in Figure 5-7. For both environments, there were six test sites containing two varieties with four replications each, or a total of 48 plots for each of the six seeding rates. Plant growth and size was much greater in the good environment than in the poor environment. The increased plant size and competition between plants resulted in fewer plants surviving to harvest, and there was little response to increased seeding rates. For this environment, the most profitable seeding rate was about 150,000 seeds per acre, which resulted in a harvest population of 110,000 plants per acre.

In the poor growth environment, there was less vegetative growth, and the reduced competition between plants resulted in increased harvest populations. Yields were much lower and generally decreased as the plant population decreased. In the poor growth environment, the most profitable seeding rate was just over 200,000 seeds per acre and resulted in a harvest population of about 180,000 plants per acre. A good rule-of-thumb for seeding rate is: Where soybean plants grow to 40 inches tall, plant 125,000 seeds per acre; where they grow to 30 inches tall, plant 175,000 seeds per acre; and where they grow to only 20 inches tall, plant 225,000 seeds per acre.

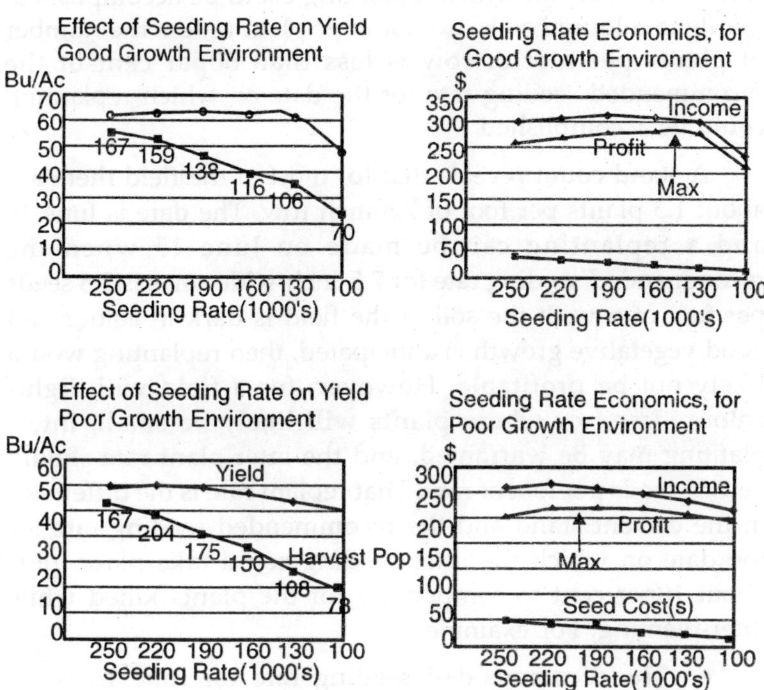

Fig. The effect of soybean seeding rate on harvest population, yield, and profit for good and poor growth environments

Replanting

Sometimes, plant stands are reduced by disease, herbicide injury, hail, insects, and flooding. If crop insurance covers the damage, consult the insurance agent before taking action. When all plants of a field are lost, it is realistic to replant if adequate growing season remains for the crop to mature. Areas of fields may be replanted while leaving the remainder of the field as is, and areas of inadequate stands can be thickened by inter-planting additional seed. If the stand loss is random or erratic throughout the field, a stand count should be taken to determine the number of plants remaining. For dark soils, do not inter-plant more seed unless the number of plants per foot of row is less than 45per cent of the recommended seeding rate for the date on which replanting could be accomplished. For light-coloured soils, do not inter-plant unless the number of plants per foot of row is less than 60per cent of the recommended seeding rate for the date on which replanting could be accomplished.

A stand count reveals that for most of the field there are about 1.5 plants per foot of 7.5-inch row. The date is June 10 and a replanting can be made on June 15 when the recommended seeding rate for 7.5-inch-wide rows is 2.8 seeds per foot of row. If the soil in the field is dark in colour and good vegetative growth is anticipated, then replanting would likely not be profitable. However, for a field with light-coloured soil or where plants will likely be small, inter-planting may be warranted, and the inter-plant rate should be 1.65 seeds per foot of row. That replant rate is the difference in the current stand and the recommended seeding rate for the date on which the inter-seeding could take place, plus about 10per cent to compensate for the plants killed while inter-planting. For example:

- The recommended seeding rate for June 15 is 2.8 seeds per foot of 7.5-inch row.
- There are currently 1.5 plants per foot of row.

  2.8 seeds per foot - 1.5 plants per foot = 1.3 plants per foot of row needed

  1.3 seeds per foot × 110per cent = 1.43 seeds per foot of row or about 100,000 seeds per acre

  or 40 pounds of seed if there were 2,500 seeds per pound.

Planting Depth

One inch to 1.5 inches is the ideal planting depth where tillage is used. Where tillage is used, the soil should be free of large clods to ensure good seed-soil contact and good seed coverage. Shallow planting (¾ to one inch) in late April promotes more rapid emergence than deeper planting. However, be aware of the increased exposure to herbicides, which may damage young seedlings. In late April soil temperatures at one-inch depth are three to eight degrees warmer than at two-inch depth. After May 15, the air temperatures are higher, and the probability of crusting increases.

It is a poor practice to plant deeper than one to 1.5 inches, because a crust may form above the seed and reduce emergence. It takes the combined pressure of many plants to break through the crust. In the process, many of the hypocotyls are broken, and the seedlings do not emerge. When planted at a one-inch depth, the seed is more likely be inside the crust layer. As the seed swells in the germination process, the soil crust is broken, and a higher percentage of plants emerge. On some crusting silt loam soils, deep planting results in 25per cent to 50per cent mortality during emergence. Where soil crusting is a problem, no-till planting and crop residue are preferred. Adequate crop residue prevents the formation of soil crust and aids in stand establishment. Three-fourths-inch to one-inch seeding depth is ideal for no-till seeding.

Fertilization Recommendations

For optimal yields on mineral soils with subsoil pH greater than 6.0, the pH range should be maintained between 6.2 and 6.8. On mineral soils with subsoil pH less than 6.0, the range should be higher (6.5 to 6.8).

Lime should be added to soybean fields when pH levels drop below the optimal range. A soil test will be necessary to calculate lime requirements. Lime may be applied anytime for recommendations of two tons or less. Fall applications will allow time for lime to raise the soil pH before spring planting. Split applications will be required for recommendations larger than four tons per acre—half before plowing and half after plowing. Regardless of the recommendation, no more than eight tons of lime should be applied in one season. Lime application rates for no-till fields should be one half of recommendations given for a tilled field sampled to an eight-inch depth.

Nitrogen (N)

Soybeans, like other legumes, have the ability to form a symbiotic relationship with nitrogen-fixing bacteria. These bacteria can fix adequate atmospheric nitrogen to produce a yield of 50 to 60 bushels per acre. Little benefit has been obtained from adding nitrogen to wellnodulated soybeans. Soybeans adjust to early-applied nitrogen by fixing less nitrogen from the atmosphere; applications after flowering have not shown a consistent or predictable yield advantage.

Soybeans also do not respond to starter nitrogen. Bacterial infection occurs soon after emergence, and nitrogen fixation begins as early as Growth Stage V2. Yield-limiting deficiencies of nitrogen are uncommon in soybeans. Deficiencies may occur temporarily during extended cool and/or wet soil conditions after planting. These short-term situations should not lower yields, and nitrogen fixation will quickly resume with warmer temperatures and drier soils. Deficiencies seldom occur later in the growing season. However, disease, such as soybean cyst nematode, or extended hot and dry weather may limit the ability of plants to absorb nutrients and produce symptoms that resemble nitrogen deficiency.

Nitrogen fertilizer may be necessary the first time soybeans are planted in a field, even when seed inoculation is used. If the crop does not have a dark green colour by early July, 75 pounds of nitrogen per acre should be applied as urea. To ensure a reliable source of inoculation in new fields, soybeans should be grown for two years, and the seed inoculated each year.

Phosphorus (P)

Soybeans require relatively large amounts of phosphorus. It is not unusual for a 60-bushel-per-acre crop to contain 48 pounds of phosphate (P_2O_5) in the grain. Although phosphorus is taken up throughout the growing season, the period of greatest demand occurs during pod development and early seed fill (Growth Stages R3 – R5). Deficient plants seldom exhibit specific leaf symptoms. Generally, phosphorus-deficient plants will be stunted, a symptom easily confused with disease and environmental stress symptoms. Plant and soil tests are the most reliable methods to ensure against phosphorus deficiency. Soil-test phosphorus levels should be maintained between 15 and 30 parts per million. Phosphate recommendations are based on the yield potential of the field and the corresponding phosphorus levels from a recent soil test.

Potassium (K)

Soybeans require large amounts of potassium. It is essential for vigourous growth, yet never becomes a part of protein molecules and other organic compounds. Potassium is not involved extensively in biological activities in the soil. Most of the total plant potassium will be in the seed at maturity (1.4 lbs per bushel). Deficiencies are not common but are easy to recognize by yellow leaf margins.

Soil Cation Exchange Capacity (CEC) may affect potassium availability, so the critical level increases as the CEC increases. The critical level for soybeans (ppm) is 75 + (2.5 × CEC). For soils low in potassium, recommendations are designed to provide more potash than crop removal, so that soils will build up above the critical level in four years. Potash should be applied annually until soil test potassium is above the critical level. Once above the critical level, recommendations are made to replace soil potassium removed by the crop. These recommendations are slightly above the critical level to account for soil sampling or analytical variation. Depending on the CEC, the range to maintain soil-test potassium levels for optimum soybean production is between 100 and 180 ppm. Potash recommendations are given. These recommendations are dependent upon a field's yield potential, CEC, and soil-test level.

Calcium (Ca) and Magnesium (Mg)

Soybeans require a minimum exchangeable soil-test level of 200 and 50 ppm (400 and 100 pounds) of calcium and magnesium, respectively. In most cases, these requirements are automatically met when soils are maintained at the proper soil pH. Soybeans will grow well over a wide range of calcium to magnesium ratios and should not need additional calcium as long as the proper pH is maintained and soil calcium levels are higher than magnesium. Soils naturally low in magnesium should be limed with dolomitic limestone. Dolomitic lime is an economical source of magnesium and still contains generous amounts of calcium.

Sulfur (S)

Soybeans use large amounts of sulfur. A crop yielding 60 bushels per acre contains about 25 pounds of sulfur, 15 pounds of which is in the grain. Soils with more than 1.0per cent organic matter usually supply adequate sulfur for high yields. Deficiencies generally occur during cool wet weather on sandy soils and/or soils low in organic matter. Soil tests are not reliable in predicting crop response to sulfur. A continuing plant-analysis programme is the best guide to confirm the need for additional sulfur. If a sulfur need is identified, several suitable materials, such as gypsum, potassium sulfate, or potassium sulfate magnesia, will correct the deficiency.

Manganese (Mn)

Even though manganese deficiency in soybeans is not a widespread problem, its occurrence is more common than the other micronutrient deficiencies. Deficiencies are more likely to occur in glacial lakebed, glacial outwash, peat, and muck soils. Soil pH is the most important factor affecting manganese availability (it becomes less soluble at higher pH levels), but other factors such organic matter, soil type, and weather may magnify the problem. On silt loams and clayey soils, manganese deficiency seldom occurs below pH 6.8. It may occur on sandy soils that are high in organic matter with a pH as low as 6.2. Muck and peat soils occasionally are deficient at a pH as low as 5.8. Pale yellow to nearly white leaves with distinct green veins is the most visual symptom of manganese deficiency. Deficiency symptoms will first appear on younger leaves. In more severe cases, the plants will become stunted.

Manganese may be banded for wide-row soybeans, but narrow rows require foliar applications. Generally, when the plants have two or three trifoliolate leaves (Growth Stages V2 or V3), a foliar application of four to eight pounds of manganese sulfate will usually correct minor deficiencies. Multiple applications may be needed when both the surface and subsoil have high pH values.

Manganese fertilizers should probably not be mixed with herbicides such as glyphosate to prevent the loss of weed control. Producers should examine the herbicide label to confirm that the product selected will not interfere with the activity of the herbicide. Spraying at the optimal time for weed control and using the manganese chelate product, EDTA, may lower the potential for antagonism between fertilizer and herbicide.

INSECT CONTROL

Insecticide application may be required at any time throughout the development of the crop. At planting time, seed treatments may be warranted for protection against seed maggots. Early-planted fields may require control of over-wintering bean leaf beetles if severe stand loss appears likely. Prior to flowering, soybeans can tolerate up to 40per cent defoliation by insects without an economic loss in yield. Soybeans become more susceptible to defoliation from flowering through pod-fill, and defoliation should not be allowed to exceed 15per cent during that time.

Once pods have set, pod injury by bean leaf beetle may be a problem. If pod injury is detected, a rescue treatment is warranted when injury of 10per cent or more of the pods appears imminent. Insecticides recommended for the control of several soybean insects are presented.

Chapter 7

Small Grain Production

The major objective of a small grain production system is the interception, fixation, and storage of sunlight energy. The most important components of such a system are variety selection, timely planting, disease control, and adequate fertilization. The effects, interactions, and relationships of various inputs to small grain production systems are discussed here.

WHEAT PRODUCTION

The production of soft red winter wheat and enjoys an outstanding reputation for the quality of its crop. Flour made from soft red winter wheat is superior for making cakes, crackers, cookies, and all sorts of pastries. Any contamination from hard red wheat or soft white wheat in marketing channels reduces its market value and the quality of flour made from it.

Attempting to produce ultra-high yields by using extra inputs is not profitable for most wheat producers. That is because the climate of limits maximum wheat productivity. When we have one of those rare dry springs with low disease levels followed by a cool June, the yields of some fields have reached 120 bushels per acre or more. Because those good growing seasons are rare, we should manage for the more normal weather. It is the weather that usually prevents us from taking advantage of high management inputs such as high seeding rates and extra nitrogen.

The most prudent production system is one of defensive management—planting after the fly-safe date to dodge diseases; holding seeding and nitrogen rates down to reduce disease and lower the cost of production; using resistant varieties instead of applying fungicides, etc. This management system will not produce the maximum possible yield in those really good years, but it will be the most profitable system for all those other years (the norm) when the weather is not ideal for maximum yields.

A research study conducted at three locations in 1999 and 2000 compared high and low levels of inputs to determine their effect on winter survival, tiller production, heads per acre, and yield. The high-management system included a seeding rate of 30 seeds per foot of row, 50 pounds of fall-applied nitrogen followed by 100 pounds in April, and the application of fungicide at various times. The low-management system used a seeding rate of 20 seeds per foot, 25 pounds of nitrogen in the fall, 75 pounds in the spring, and a disease-resistant variety that required no fungicide. The high-management system cost $31 per acre more than the low-cost system but produced only 2.7 (89.2 vs. 91.9) more bushels of grain per acre. The extra seed and nitrogen alone raised cost by $16 per acre.

High yields and low cost of production are necessary for wheat to be a viable economic partner in the crop-rotation sequence. Increased profitability will only come from improved management. The guidelines presented here will help minimize the factors limiting wheat yields and also lower production costs.

Variety Selection

The yield potential of currently available varieties is generally in excess of 150 bushels per acre. This yield is not approached, however, primarily because of a short grain-fill period caused by high air temperatures in late June and early July which kill the crop. Accordingly, growers need to select wheat varieties with high yield potential, high test weight, good winter hardiness, and good straw strength. Always select varieties with adequate resistance to the diseases prevalent in your area of the state. Information on variety performance should be obtained from multiple sources, such as seed companies and university performance trials, where multiple sites and years of testing are presented.

Always plant more than one variety each year to reduce the risk of disease losses and to spread out harvest dates. Select varieties with resistance to wheat spindle streak mosaic, powdery mildew, and leaf rust. Varieties with moderate resistance to *Stagonospora* blotch and Fusarium head scab are also available. Avoid varieties highly susceptible to Fusarium head scab.

High-Quality Seed and Seed Treatment

Purchase only high-quality seed that has been thoroughly cleaned to remove shriveled kernels and that has a germination of 90per cent or better. All seed should be treated with a seed-treatment fungicide to control seed-borne diseases like loose smut, common bunt, Fusarium scab, and *Stagonospora* glume blotch. Treatments containing the fungicides difenconazole or tebuconazole (TBZ) have been very effective in controlling most seed-borne diseases. TBZ can be added to improve efficacy against seed-borne Fusarium found on scab-affected seed.

Crop Rotation

Plant wheat following soybeans. A three-year rotation of corn-soybean-wheat appears to be optimum for sustained yield of all three crops. Crop rotation is the most effective method to reduce pathogen populations that affect the three crops in the sequence. The purpose is to provide enough time away from the host plant for pathogens to die out before that crop is planted again. Wheat should never follow wheat or spelt in the rotation sequence.

Soil-borne diseases, like Take-all and *Cephalosporium* stripe, can cause complete crop failure in non-rotated fields. Foliar diseases, like powdery mildew and *Stagonospora* glume blotch, may also become more of a problem. Wheat should not follow corn in the rotation because the same fungus that causes Gibberella stalk rot in corn also causes Fusarium head scab in the wheat. Planting wheat into corn residues greatly increases the risk of a severe outbreak of scab in the wheat crop. Wheat also serves as an excellent rotation crop for corn and soybeans, allowing populations of pathogens (like soybean cyst nematode and Sclerotinia) to decline before host crops are again planted in the field.

Land Selection and Preparation

Wheat grows well in a range of soil types; however, well-drained soils with medium to fine texture produce the highest yields. Adequate drainage is essential; thus, tiling poorly drained fields is important. Plan the crop-rotation sequence far enough in advance to plant early-maturing soybean varieties in fields to be planted to wheat in the fall. This will permit planting of wheat at the optimum time for maximum winter survival and yield potential. Drilled, medium-season soybean varieties, planted early yield as well as full-season varieties.

Planting no-till wheat into soybean stubble has been very successful in reducing erosion and almost totally eliminates spring heaving, which also reduces production costs. Soybean residues should be evenly spread across the field during harvest to ensure uniform seeding depth (1.5 inches). Do not plant into soils that are too wet and monitor planting depth when the soil is hard and dry.

Planting Date

Wheat should never be seeded prior to the fly-safe date because of the possibility of severe damage by diseases and Hessian fly. The best time for seeding is a 10-day period starting the day after the fly-safe date. Long-term average yields are highest from seedings made during that time. Seeding during that time usually produces ample growth for winter survival and reduces the likelihood of fall disease infections and attack by potentially damaging insects. Occasionally, when freezing weather is delayed until late November or early December, wheat seeded more than three weeks after the fly-safe date is equal in yield to that seeded during normal planting time. Because of reduced fall growth, late seeded wheat is less winter hardy and more susceptible to spring heaving.

Seeding Practices

When planting at the proper time and into soil that is not too wet, seed should be planted 1.5-inches deep. Row width should be six to eight inches. Planting by bushels per acre is very inaccurate due to variability in seed size from year to year and from one variety to another. Low seeding rates result in inadequate stands and winter injury, while excessively high rates increase lodging and disease pressure.

Do not plant weak-strawed varieties prone to lodging. Calibrate the drill each year for each variety and seed lot planted. The optimum seeding rate is 1.2 to 1.6 million seeds per acre (18 to 24 seeds per foot of 7.5-inch row when planting during the two weeks following the fly-safe date). During the third and fourth week after the fly-safe date, plant 1.6 to 2.0 million seeds per acre (24 to 30 seeds per foot of row). Do not plant faster than the speed at which the drill was calibrated. The number of seeds per pound and germination rates are critical factors that need to be known before a proper seeding rate can be determined and the drill calibrated. This information should be listed on the bag of seed. The information in Tables 6-2 and 6-3 can be used to accurately calibrate grain drills.

Row Width

Wheat row spacing work conducted in the mid-1980s indicated that wheat grown in 14-inch rows produced yields that were 94per cent of yields from rows spaced seven inches apart (61 bushels compared to 57.3 bushels). Because the seeding rate per foot of row for wheat is the same for all row widths, the seed cost for 15-inch rows is half that for 7.5-inch rows. When wheat seed costs $12 per unit and wheat grain is worth $3 per unit, the lower yield from wide rows is almost offset by the reduced seed cost. The additional savings from drills with fewer seed meters and planting units made the two row spacings equally profitable based on the 1980s data.

The results of studies conducted in 2000 and 2001 appear in Tables 6-4 and 6-5. In both years, the plots were planted within 10 days after the fly-safe date at the rate of 25 seeds per foot of row for both row spacings (120 lbs. and 60 lbs. for 7.5-inch and 15-inch rows, respectively). Nitrogen (30 lbs. per acre) was applied at planting each year to stimulate fall growth, tillering, and improve winter hardiness.

Table: Effect of Wheat Row Spacing on Plant Height, Test Weight, and Yield for 18 Varieties and Six Test Sites in 2000.

	7.5 In. Rows	15 In. Rows	Difference
Height	37.0	36.7	0.3
Test Wt.	54.9	53.9	1.0
Yield	60.6	56.2	4.4

Table: Effect of Wheat Row Spacing on Plant Height, Test Weight, and Yield for 22 Varieties and Two Test Sites in 2001.

	7.5 In. Rows	15 In. Rows	Difference*
Height (in)	38.0	37.0	1.0
Test Wt. (lbs/bu)	55.6	55.6	0.0
Yield (bu/ac)	72.9	70.7	2.2

* Three varieties produced more yield in 15-inch rows than in 7.5-inch rows

These data indicate that some wheat varieties may be more profitable to produce in wide rows than narrow rows due to savings in seed and machinery cost. Varieties that perform well in wide rows tend to be either tall by nature or grow tall due to favorable weather. They also have a non-erect growth habit that allows them to fill in the wide-row middles. High rates of tillering also favor higher yields in wide rows. Tillering is favored by planting within seven days after the fly-safe date and the application of 30 pounds of nitrogen at planting. Normally, 15-inch-row wheat yields 5per cent to 15per cent less than wheat grown in 7.5-inch rows. In 2001, excessive tillering and vegetative growth reduced that normal difference in yield.

Lodging Control

Lodging is a serious deterrent to high yields. Cultural practices that tend to increase grain yield also increase the likelihood of lodging. Using recommended seeding rates (18 to 24 seeds per foot of 7.5-inch row), applying proper rates of nitrogen, and selecting lodging-resistant varieties prevents lodging in high-yield environments where yields of 100 bushels per acre are anticipated. When lodging occurs, the severity of foliar disease increases, resulting in reduced grain yield and quality. Additional effects of lodging are reduced straw quality and slowed harvest. The prevention of lodging increases dividends through a combination of reduced input costs and improved grain and straw quality.

Fertilization

Nitrogen

Providing adequate nitrogen for the wheat crop is an important step toward high yields. However, as the nitrogen rate increases, the potential for lodging and disease also increases. The amount of fertilizer nitrogen required varies greatly, depending on the level of soil organic matter, carry-over nitrogen from previous crops, and yield goal. In general, each 1per cent of organic matter supplies eight to 12 pounds of nitrogen per acre through organic matter oxidation. Nitrogen available from a previous soybean crop ranges from 20 to 40 pounds per acre.

Current recommendations are to apply 20 to 30 pounds of nitrogen at planting to stimulate fall vegetative growth. Spring nitrogen should be applied between March 1 and April 30. When starter nitrogen is applied in the fall, spring applications in mid- to late April are satisfactory. When no nitrogen is applied in the fall, the spring application should be made before April 15.

Most sources of nitrogen are satisfactory for wheat. Urea and 28per cent solutions (urea-ammonium nitrate) are often the most common. Urea has the least potential to cause damage to the crop. Damage is generally insignificant from broadcast applications of 28per cent solution applied early, but the potential for damage increases as the crop matures. Dribble band applications will minimize damage from 28per cent solutions. Urea-ammonium nitrate solutions will have some nitrogen available immediately at application time; urea will have a short lag as it converts to ammonium and nitrate forms of nitrogen. The nitrate fraction of 28per cent solutions will be susceptible to loss after application until plant uptake. Urea may have volatilization losses if temperatures are exceedingly warm.

The spring application should be the total nitrogen recommendation less the amount applied in the fall. No credits are given for previous crops. For example: A 90-bushels-per-acre wheat crop would require 110 pounds of nitrogen. If the grower applied 30 pounds in the fall, the remaining 80 pounds should be applied in the spring.

Table: Nitrogen Recommendations for Wheat.

Yield Potential (bu per acre)	Nitrogen Rate (lb per acre)
60	60
70	75
80	90
90	110
100	130

A split spring application programme may be a benefit in poorly drained fields that are prone to nitrogen loss or in fields prone to lodging. For these programmes, it is important that the first application occurs soon after initial green-up and the second application at initial jointing (Feekes Growth Stage 6). The time of application is not as critical in a single spring topdress (providing some nitrogen was applied in the fall), but applications should be made after initial green-up and before the second visible node on the stem. Up to half the total intended amount of nitrogen can be applied in the second application.

Phosphorus

Small grains respond well to phosphorus fertilizer on soils testing up to 45 ppm. Therefore, to maximize yields, the soil P level (Bray P1) should be 45 ppm or higher. A 1:4:2 ratio provides the proper balance of starter fertilizer. The recommended phosphorus applications. Phosphorus should be applied before planting when the soil-test level is below 45 ppm. Actual phosphorus recommendations are determined by the yield goal and soil-test level. Phosphorus and fall-applied nitrogen are often applied as diammonium phosphate (DAP) or monoammonium phosphate (MAP).

Potassium

The small-grain response to potassium application is less than that of phosphorus. When the K soil-test level is maintained at 75 ppm plus five times the CEC (Cation Exchange Capacity), a potassium application equal to the crop removal plus 20 pounds of K2O per acre produces optimum yields if no other factors are limiting. If the soil-test K level is maintained at 175 ppm plus five times the CEC, then no application of potassium is needed for optimum yields.

Recommendations for potash are based upon the yield goal, soil Cation Exchange Capacity (CEC), and the soil-test level. Soils with higher CEC values have a greater chance of potassium becoming unavailable to the crop and require more potash than low CEC soils.

Manganese

Wheat is almost as sensitive to manganese deficiencies as is soybean, and the problem occurs in the same areas of fields. Deficiency symptoms are usually not severe enough to be seen but will reduce yield. Maintaining soil pH between 6.5 and 7.0 will usually eliminate the problem. Manganese sulfate can be applied in a band or in contact with seeds when it is recommended by a soil test. A more practical application method is to mix some manganese sulfate with liquid nitrogen for application at or before green-up in the spring.

Disease Control

Yield losses as high as 30 to 50per cent are common where no disease control is practiced. Effective disease management requires knowledge of the diseases most likely to occur in a production area. Producers should fine tune their disease-control strategies for those few diseases encountered each year.

A comprehensive wheat disease-management programme consists of the following practices:

Select varieties with resistance to the important diseases in the area. *Stagonospora* glume blotch is most severe in central, west central, northwestern, and southern region. Monitoring wheat diseases aids a producer in selecting varieties with resistance to the common diseases of his or her community.

Plant well-cleaned, disease-free seed, treated with a fungicide that controls seeding blights, bunt, and loose smut.

Plant in a well-prepared seedbed, after the fly-safe date.

Rotate crops; never plant wheat where the previous crop was wheat or spelt. A two- to three-year rotation from wheat prevents most pathogens from surviving in fields.

Plow down residues from heavily diseased fields, especially those affected by *Cephalosporium* stripe or Take-All. Plowing enhances decomposition of residue and death of the disease-causing fungi.

Use a well-balanced fertility programme based on a soil test. Apply sufficient amounts of phosphorus, nitrogen, and potassium in the fall for vigourous root and seeding growth. Spring top-dress with nitrogen at the rate recommended to achieve the yield goal. Excessive nitrogen increases the severity of foliar disease and lodging.

Control grass weeds. Destroying volunteer wheat, quackgrass, and other grass weeds in and around potential wheat fields reduces the amount of disease inoculum available to infect the crop.

Apply fungicides only if warranted. Scout fields from flag leaf emergence through flowering. Foliar fungicides are able to control the following diseases—powdery mildew, leaf rust, and *Stagonospora nodorum* leaf and glume blotch. *Stagonospora* leaf blotch and powdery mildew are the most severe. Know symptoms, severity ratings, and disease thresholds before scouting fields. The earlier the fungicide is applied after reaching the threshold level, the greater the yield response.

Planting disease-resistant varieties is the most effective and economical means for controlling diseases. Select resistant varieties based on research conducted by universities and seed companies. Varieties are available with moderate to high levels of resistance to leaf rust, powdery mildew, and wheat spindle streak mosaic. Varieties with moderate levels of resistance to *Stagonospora* leaf and glume blotch and *Fusarium* head scab are also available. However, varieties rarely have good resistance to all diseases. When varieties have high resistance to a disease, they effectively limit losses in yield. Resistance to leaf rust and powdery mildew may fail due to the development of new races of the pathogens. Combining the use of resistant varieties with good crop rotations, planting after the Hessian fly-free date, and the use of seed-treatment fungicides will improve disease control.

Scouting fields for disease is essential when growing moderately susceptible and susceptible varieties to determine the need for fungicide applications. This involves checking the level of disease on 30 to 50 individual tillers randomly selected throughout the field. Fields should be scouted for powdery mildew at flag-leaf emergence and boot stage and scout for *Stagonospora* leaf blotch and leaf rust at boot stage and full-head emergence.

The upper two leaves on tillers and the glumes on heads contribute most to grain fill. Thus, it is important to keep these upper plant parts free of disease to avoid yield loss. Fungicides should be used only on susceptible and moderately susceptible varieties in fields that have a yield potential of 60 or more bushels per acre. Disease thresholds are 1per cent of leaf area affected on the leaf below the flag leaf up to boot stage (Feekes Growth Stage 8 through 10), and 1per cent of leaf area affected on the flag leaf between head emergence and flowering. When these disease levels are present, a fungicide should be applied as soon as possible to protect leaf tissue before more becomes infected. One percent leaf area affected roughly translates to five to 10 leaf-rust pustules, two to three powdery-mildew pustules, or one to two *Stagonospora nodorum* blotches.

Insect Control

Several different insects can be important on wheat. Management of insect pests affecting wheat often emphasizes nonchemical control measures. Hessian fly is controlled primarily by delaying planting until late September or early October, depending on location of the field in the state. Cereal leaf beetle and aphids are usually controlled by beneficial parasites. However, populations of some pests, especially cereal leaf beetle, armyworm, and sometimes aphids, may occur in numbers warranting rescue treatment with insecticides. The insect pests that may impact wheat are reviewed here:

The Hessian fly:[BOLD] It passes through two generations per year in which adult flies deposit eggs, maggots hatch on leaves and feed on stems, and then maggots pupate into the commonly recognized flaxseed stage. Damage by the maggots occurs in the late spring and early fall following activity by adults in early spring and late summer. Flaxseed pupae are located within the leaf sheaths of plants in the spring, resulting in the broken wheat stems and lodging associated with that damage. Under serious infestations, the problem is generally detected after the damage has been done and the fly is in the flaxseed stage. Thus, rescue treatments are not warranted. The major tactic for controlling Hessian fly is planting wheat after the Hessian fly-safe date for your county.

English grain aphid and the cherry oat aphid: They may cause limited feeding injury. The greenbug, which produces a toxin that affects the wheat plant. To determine the need for treatment, first identify the aphid. English grain aphid has black cornicles or tailpipes on the tip of the abdomen, oat-bird cherry aphid has a red-orange spot between the cornicles, and the greenbug has a dark green stripe on the back and the tips of the cornicles are black. Keep in mind that natural predators usually control most aphid populations on small grains. Aphids also are important because they may transmit the barley yellow dwarf virus that causes stunting and yellowing of wheat and other small grains. However, it is not economically feasible to control transmission of barley yellow dwarf virus with insecticides because aphids can transmit the virus within six hours of landing on the plant.

Cereal leaf beetles: They appear in the spring and lay eggs, which hatch into larvae that feed on wheat and oat leaves. Larvae appear as small black slugs due to accumulated fecal matter on their backs. There is one generation per year with new adults appearing in late spring. A complex of parasitic wasps generally controls cereal leaf beetle, but treatment of fields may be warranted when mild winters adversely affect natural control. An infestation averaging one larva per stem may result in a loss of three bushels per acre.

Adult armyworms: They become active in late April and early May and are attracted to grass crops, including wheat. Larvae are active in late May and June and can feed on leaves and emerging heads. Most serious damage occurs when larvae feed on stems and clip heads completely off. Detection of larvae is initially along the edge of fields and low-lying areas. When six or more larvae can be seen per linear foot of row, or head clipping is evident and larvae are not fully grown, a rescue treatment may be needed.

Weed Control

Wheat competes well with weeds, especially when good production techniques result in an initial uniform stand establishment and when loss of stand due to winter injury is minimal. Effective weed control and prevention of weed seed production in prior crops will reduce the risk of weed problems in wheat. Some wheat fields can benefit greatly from herbicide application, and failure to scout fields and take the appropriate measures can result in yield loss and harvesting problems in these fields. The weeds that appear above the wheat canopy late in the season, such as ragweed and Canada thistle, can often be easily controlled with a spring herbicide treatment. The most common weed problems in wheat are:

1. Winter annual weeds, such as common chickweed, purple deadnettle, shepherdspurse, and field pennycress. These weeds become established in the fall along with the wheat and can interfere with early development of wheat in the spring. Dense populations of winter annual weeds should be controlled in late fall or early spring to minimize interference with wheat growth.
2. Wild garlic, which contaminates harvested grain with its bulblets. Several herbicides are effective if applied in the spring after garlic has several inches of new growth.
3. Canada thistle, which can greatly suppress wheat growth due to its tendency to occur in dense patches. Most wheat herbicides have some activity on thistle and can suppress it adequately, if not applied too early in spring.
4. Summer annual broadleaf weeds, such as common and giant ragweed, which can begin to emerge in late March. A healthy wheat crop can adequately suppress these weeds, but herbicide application is occasionally warranted.

It is very important to apply herbicides at the correct stage of growth (Figure 6-4) of the wheat plants in order to avoid herbicide injury to the wheat. When wheat has not yet reached the jointing stage, any herbicide labeled for wheat can be safely applied. As the wheat growth stage advances past jointing and then past boot stage, herbicide choices become much more limited. Most herbicides can be applied in nitrogen fertilizer solution when the wheat is top-dressed. This may increase injury somewhat, and some labels recommend adjusting surfactant rates to minimize injury.

PRODUCTION OF OTHER SMALL GRAIN SPECIES

Fertilizer recommendations for the other small grain species are provided in Tables.

Spelt

Spelt, sometimes referred to as Speltz, is a type of wheat not considered in the official grain standards. Spelt is primarily used as part of the grain ration for livestock and is usually ground or milled before feeding. It is an excellent replacement for oats in rations needing bulkiness. Spelt can be grown in areas where winter wheat safely survives the winter. Common spelt is very susceptible to leaf rust, powdery mildew, stinking smut, and loose smut. Spelt is tall, with moderately weak straw, and late maturing compared with most wheat varieties. Because spelt does not thrash completely free of its surrounding chaff, there is no official bushel weight. Usually, 20per cent to 30per cent of its total weight is the surrounding chaff. This should be considered when comparing yields with wheat. Treatment of the seed with a fungicide prior to planting helps prevent a disease problem, but the fungicide label should be consulted for clearance before use on spelt. Spelt has similar susceptibility to insects as wheat.

Currently available varieties include Champ, Comet, Oberkulmer, and Sava. Spelt should be managed like wheat except for the seeding rate (15 to 20 seeds per foot of 7.5-inch row or two to three bushels per acre) and nitrogen application rate of 50per cent to 70per cent that of wheat. Spelt should not be planted following wheat, and the fly-safe date should be adhered to. Mature grain standing in the field dries more rapidly after rain than the other small grains because of the water shedding characteristics of the grain chaff. Combine settings include a slow cylinder speed (similar to soybeans) and very little air to the screens.

Spring Oats

Several varieties of spring oats (Armor, Burton, Jay, Ogle, Ida, and Gooding) having high yield potentials, good test weight, and stiff medium-short straw are available. Some varieties are resistant to the common diseases of oats. Spring oats is the first crop to be planted in the spring. Therefore, the selection of fields with well-drained soil is essential to permit timely planting. Spring oats should be seeded as early in the spring as soil conditions permit, preferably between March 1 and April 15. Grain yields decrease rapidly as seeding is delayed past mid-April. The proper seeding rate is 15 to 20 seeds per foot of 7.5-inch row (75 to 100 pounds of high-quality seed per acre). The seed should be planted no more than one inch deep to assure rapid emergence. Although oats can be established using no-till seeding techniques, little if any crop residue should be present to allow the soil to warm rapidly and aid germination and emergence. Fall preparation of the seedbed eliminates the need for tillage in the spring and sometimes permits earlier planting than when tillage is needed. This technique also eliminates the soil compaction associated with soil preparation.

Soil pH should be above 6.0 unless a legume is also seeded, in which case the soil pH should be 6.5 to 7.0. Phosphate and potash should be applied in the fall. Nitrogen should be applied in the spring anytime before emergence.

Winter Barley

Although barley is an excellent animal feed and easily replaces corn in rations, very little winter barley is grown because of a general lack of winter hardiness. If grown, barley should be seeded between September 15 and 30 to increase its chances of surviving the winter. Seed 15 to 20 seeds per foot of 7.5-inch row (90 to 120 pounds of high-quality seed per acre).The soil pH should be between 6.5 and 7.0, and recommended levels of phosphorous will improve winter survival. Twenty pounds of nitrogen per acre and the recommended phosphate and potash should be applied preplant.

Spring Barley

Yield performance of spring barley, and it is an unprofitable crop to produce.

Triticale

Triticale is a human-made species resulting from a cross of wheat (Triticum) and rye (Secale). Hybrids of wheat and rye crosses date back to 1875, but intensive work on such hybrids was started only about 30 years ago at the University of Manitoba. Both winter and spring types have been developed. The protein content of triticale is normally higher than wheat, and the amino acid composition of the protein is similar to wheat. Feeding trials have found triticale to be unsatisfactory for hogs, and triticale produced less weight gain and feed efficiency in beef cattle than barley. Forage yields have been similar to wheat and rye. The cultural practices and fertilization practices recommended for wheat are satisfactory for triticale.

Winter Rye

Winter rye is usually used as a winter cover crop because of its tolerance to adverse growing environments. Winter rye is the most winter-hardy and earliest-maturing cereal grain grown. It is more productive than other cereals on infertile, sandy, or acidic soils, and on poorly prepared land. When used as winter cover, winter rye is usually seeded at a rate of 60 to 90 pounds per acre. When grown for grain production, winter rye should be seeded between September 20 and October 20 at a rate of 85 to 115 pounds of high-quality seed per acre. The seed should be planted one inch deep. When used as a winter cover crop or a green manure crop, it should be seeded in early September. Fertilization is similar to wheat, but nitrogen application should be limited to 50 pounds per acre. Rye competes well with weeds, and herbicides are generally not needed. Bromoxynil (Buctril), MCPA, and 2,4-D are labeled for use.

WEED CONTROL FOR SMALL GRAINS

Winter small grains usually compete quite well with most weeds, especially when sound management practices are used to produce a good stand. Most broadleaf weeds that do become a problem can be controlled with herbicides. These herbicides reduce small grain yields if applied at the incorrect stage of growth. Grasses usually do not become an economic problem in small grains, and at the present time, no herbicides are available to control grasses if a problem with grasses occurs. Small grains are very sensitive to triazine residues. To avoid an atrazine or simazine residue problem, do not use either of these products for weed control in corn when a small grain crop will be seeded that fall and no more than 0.8 pound active ingredient per acre (1 lb per acre 80WP, 0.9 lb per acre of Nine-O, or 0.8 qt per acre of 4L) when oats will be planted the following spring.

Chapter 8

Forage Production

The forage industry plays a major role in agriculture. Forages are environmentally friendly. They protect soils from erosion, improve soil tilth, help reduce pesticide use, and enhance agricultural profitability. Forages are vital to agriculture, protect our soil and water resources, and add beauty to the state. All forage crops respond dramatically to good management practices. Higher yields, better forage quality, and improved persistence result from paying attention to the basics of good forage management. The objective of this guide is to help producers achieve the high potential of forages grown.

PERENNIAL FORAGES

Species Selection

The selection of forages for hay, pasture, and conservation is an important decision requiring knowledge of agronomic characteristics, forage species adaptation to site and soil characteristics, and potential feeding value of forage plants. The intended use of forages, dry matter and nutritional requirements of livestock to be fed, seasonal feed needs, harvest and storage capabilities, and seasonal labor availability influence which species to grow.

Agronomic Adaptation and Intended Use

The choice of species is limited to those adapted to the soils on the farm, the soil adaptation factors should be studied carefully.

Keep in mind that soil pH can be increased with lime; soil fertility can be improved with fertilizers and manure; and soil drainage can be improved with tiling. Soil drainage is usually the most difficult soil characteristic to modify.

The brief discussion of individual species presented here will also help in determining which species are best suited to a particular enterprise. Table gives commonly recommended species to consider for general soil fertility classes and forage utilization methods. The information here is by no means exhaustive, but it does cover the most common forages adapted to conditions.

Pure Stands vs. Mixtures

The decision to establish a pure stand or a mixture should be made before deciding which species to plant. Advantages of pure grass or legume stands are simpler management, more herbicide options, and greater forage-quality potential. Pure legume stands decline in forage quality more slowly with advancing maturity than do grasses, providing a wider window of opportunity for harvesting good-quality forage.

Legume-grass mixtures are common and are generally more satisfactory for pastures than are pure stands. Grass-legume mixtures are often higher yielding and have more uniform seasonal production than pure stands. Including legumes in a mixture reduces the need for nitrogen fertilizer, improves forage quality and animal performance, and reduces the potential for nitrate poisoning and grass tetany compared with pure grass stands. Including grasses in a mixture usually lengthens the life of a stand because they persist longer and are more tolerant of mismanagement than legumes. Grasses reduce the incidence of bloat, improve hay drying, are usually more tolerant of lower fertility, and compete with weeds more than legumes. The fibrous root system of grasses helps control erosion on steep slopes and reduces legume heaving. Growing grasses and legumes together often reduces the losses from insect and disease pests.

Mixtures for Hay and Silage

Keep mixtures relatively simple for hay or silage use; two to four species are usually sufficient. Hay and silage cutting schedules are easier to manage with simple mixtures. Select grass and legume species to build simple mixtures for hay or silage using the following criteria:

Rate of Establishment

Combine species with fairly similar seedling aggressiveness. Persistent plant species are often the least competitive in the seedling stage. Excessive seedling competition in a shotgun mixture can prevent persistent and desirable species from becoming established. An exception to this rule is the use of companion crops, or fast-establishing short-lived perennial or annual species used to achieve quick ground cover. Small grains and annual or perennial ryegrass are often used for this purpose. Keep seeding rates of these temporary companions low to avoid excessive competition with the slower-establishing perennial species

Time of Maturity

Species and varieties in a mixture should mature at about the same time and match your intended harvest schedule. There is considerable variation in maturity of grass species and varieties. Such information is often collected in variety testing trials.

Management Compatibility

Select species that are well adapted to the intended management. For example, orchardgrass is compatible with alfalfa on a four-cut schedule because it regrows quickly, while timothy and smooth bromegrass are compatible with alfalfa on a more lenient three-cut schedule.

Summer Production

Alfalfa produces very well during the summer months while birdsfoot trefoil and red clover generally produce less summer yield. Of the grasses, orchardgrass and reed canarygrass produce the best summer growth. Smooth bromegrass produces moderate to light summer aftermath, and timothy, meadow fescue, and perennial ryegrass are usually low yielding in the summer months. Moisture and temperature conditions affect aftermath production of cool-season grasses more than alfalfa.

Variety Performance

Use variety testing data to select species and varieties that have stable yield performance over multiple locations and years, which demonstrates good adaptation to a wide range of conditions. Performance over years demonstrates yield persistence with advancing stand age and is especially important for long rotations. Disease and pest resistance

Select species and varieties with resistance to key insects and diseases for your soils. For example, Phytophthora root rot and Fusarium wilt resistance in alfalfa are very important on soils with sub-optimal drainage, while potato leafhopper resistant alfalfa.

Forage Quality

Varieties with improved forage quality are available in some species, such as alfalfa. If high forage quality is very important, then select varieties based on this trait.

Mixtures for Pastures

Simple mixtures are desirable for hay and silage management, but recent studies demonstrate that complex mixtures of six or more species provide greater stability of forage production under grazing. Soil and environmental variability is usually high in pastures and, therefore, it is more difficult to predict which species will perform best across the variable landscape. Species dominance and spatial distribution in a pasture will be affected by variability in fertility, soil drainage, slope aspect (north or south facing), and animal traffic and grazing patterns, among other factors that influence the microenvironment. In addition, species vary in productivity during different seasons, i.e., between spring and summer grazing periods. Select grasses and legumes that fit the general soil conditions and management characteristics and that are not drastically different in palatability.

Seeding Rates

Seeding rates for individual species in pure stands and for mixtures. Seeding rate recommendations are related to seed size, germination, seedling and established plant vigour, spreading characteristics, and mature plant size. For example, more seeds per square foot are recommended for species with low seedling vigour and smaller mature plant size in order to improve establishment success and competitiveness of that species in a mixture or against weed encroachment. Increasing seeding rates above the recommended levels does not compensate for poor seedbed preparation or improper seeding methods.

There is no reliable way to predict that a specific proportion sown will result in an equivalent proportion of established plants in a mixed species seeding. The seeding rates shown for mixtures are simply varying percentages of the pure stand seeding rate recommendation. Use your best judgment to adjust the seeding rate for each species based on the relative proportion desired of that species in the mixture. Complex mixtures will often result in a higher overall seeding rate (in seeds per square foot) than simpler mixtures. This is simply a function of having more component species, each one seeded above a minimum level, to provide an opportunity for it to establish and compete in the microenvironments where it is best adapted.

Characteristics of Perennial Cool-Season Forages

Alfalfa

Alfalfa is grown on about one-half of the total hay acres. Where adapted, it is unmatched by any other forage as high-quality feed for livestock and as a cash crop. Alfalfa requires deep, well-drained soils with near-neutral pH (6.5 to 7.0) and high fertility for high yields. It should not be grown on soils with moderate to poor drainage. Alfalfa is best adapted to hay or silage harvest management. While it can be used in rotationally grazed pastures, it normally lacks persistence in permanent pastures compared with other legumes. Like most legumes, it can cause bloat. Alfalfa has good seedling vigour, excellent drought tolerance, and produces very well in the summer. Important insect pests on alfalfa include the alfalfa weevil and potato leafhopper.

To capitalize on alfalfa's potential, select newer high-yielding alfalfa varieties with adequate winter hardiness and resistance to important diseases. Recent improvements in alfalfa varieties include selection for multiple disease resistance, forage quality, and high levels of resistance to potato leafhopper.

Alsike Clover

Alsike clover is a short-lived perennial legume that is tolerant of wet, acidic soils. Alsike tolerates soils with a pH as low as 5.0, which is too acidic for red clover and alfalfa. Alsike also grows better than red clover on alkaline (high pH) soils. Alsike tolerates flooding better than other legumes, making it well suited for low-lying fields with poor drainage. Alsike can withstand spring flooding for several weeks. A cool and moist environment is ideal for alsike clover growth; it has poor heat and drought tolerance, thus usually produces only one crop of hay per year. It is susceptible to the same diseases that attack red and white clovers. Its growth habit is intermediate between red and white clover. Alsike clover must be allowed to reseed to maintain its presence in pastures; otherwise it will last only about two years. Alsike clover has good palatability, but it can cause bloat and photosensitization in grazing animals.

Birdsfoot Trefoil (**Lotus Corniculatus L.**)

Birdsfoot trefoil is a deep-rooted perennial legume that is best adapted to northern region. Birdsfoot trefoil is tolerant of low-pH soils (as low as pH 5.0), moderate to somewhat poor soil drainage, marginal fertility, and soils with fragipans. Birdsfoot trefoil can withstand several weeks of flooding and tolerates periods of moderate drought and heat. Birdsfoot trefoil has poor seedling vigour and is slow to establish. Early spring seedings are generally more successful than late summer seedings. It is subject to invasion by weeds when grown in pure stands; therefore, it is best seeded with a grass companion. It produces excellent quality forage, has fair palatability, stockpiles well, and unlike most forage legumes, it is non-bloating. Birdsfoot trefoil should be managed to allow for reseeding, which will help maintain its presence in forage stands. It is intolerant of close cutting or grazing, has slow recovery after hay harvest, and is susceptible to root and crown rot diseases.

Empire-type varieties have prostrate growth and fine stems, making them better suited to grazing. European-type varieties are more erect, establish faster, and regrow faster after harvest. Thus, they are better suited to hay production and rotational grazing. Most of the newer varieties are intermediate with semi-erect to erect growth habit.

Kura Clover (**Trifolium Ambiguum Bieb.**)

It is a perennial clover originating in Caucasian Russia. Kura clover has rhizomes, underground modified stems that spread new plants into empty spaces. It is a very hardy legume that is best adapted to long-term permanent pastures. It will tolerate continuous stocking and is very persistent once established. It has an extensive root system. In university tests, stand density improved over time. Superior animal performance on rotationally stocked mixed grass pastures containing Kura clover was associated with greater forage yield and nutritive value compared to red clover/grass pastures.

Kura clover tolerates low fertility, soil acidity, wet soils, and moderate flooding. Although it stops growing during a drought, its deep root system ensures excellent drought survival. Its major limitation is poor seedling vigour and slow establishment. It requires a specific Rhizobia inoculant. The forage grazed by livestock is primarily leaves (like white clover) and is very high in nutritive value. If sufficient grass is not present, it can induce bloat in ruminants. Its low-growing growth habit and high moisture content make it difficult to use for hay.

The first variety of Kura clover released in the United States was 'Rhizo.' Other varieties have followed, such as 'Cossack' and 'Endura.'

Red Clover (Trifolium Pratense L.)

Red clover is a short-lived perennial legume grown for hay, silage, pasture, and green manure. Red clover is better adapted than alfalfa to soils that are somewhat poorly drained and slightly acidic; however, greatest production occurs on well-drained soils with high water-holding capacity and pH above 6.0. Red clover is not as productive as alfalfa in the summer. It has good seedling vigour and is one of the easiest legumes to establish using no-till interseeding or frost-seeding techniques. Red clover is often difficult to dry for hay storage. Harvesting for silage or including a grass in the stand helps overcome this problem. When grazed, red clover can cause bloat in cattle if sufficient grass is not present.

Medium red clover varieties can be harvested three to four times per year. Mammoth red clover is late to flower and is considered a single-cut clover because the majority of its growth occurs in the spring. Most of the improved varieties are medium types and have good levels of disease resistance to northern and southern anthracnose and powdery mildew. Several new medium red clover varieties have demonstrated good stand persistence for three or even four years in university trials.

White Clover (**Trifolium repens L.**)

White clover is a low-growing, short-lived perennial legume that is well suited for pastures. It can cause bloat in cattle if sufficient grass is not present for grazing. White clover improves forage quality of grass pastures and reduces the need for nitrogen fertilizer. White clover can be frost seeded or no-till seeded into existing grass pastures. It spreads by stolons. It has a shallow root system, so it does not tolerate prolonged dry spells and usually has lower summer growth. Although well-drained soils improve production, white clover tolerates periods of poor drainage. It can be managed for reseeding to improve persistence in pastures.

Large white clover types, also known as Ladino clovers, are more productive than the smaller type White Dutch or wild white clovers. Wild white clovers persist better under heavy, continuous grazing because they are prolific reseeders. Purchase seed of stated quality to be certain of obtaining pure seed of the white clover variety desired.

Crownvetch

Crownvetch is a long-lived perennial that is most commonly used for soil erosion control, beautification, highway embankments, mine spoil areas, and other disturbed areas. It spreads by creeping underground rootstocks and has a deep taproot and numerous lateral roots. Crownvetch is best adapted to well-drained, fertile soils with a pH of 6.0 or above, although it is somewhat tolerant of moderately acid and infertile soils. It has slow seedling development, and seedlings are very sensitive to competition, so do not use a companion crop and provide good weed control during establishment. If planted with grasses, use less competitive grasses such as timothy or Kentucky bluegrass rather than orchardgrass or tall fescue. Animals do not readily accept crownvetch forage and must be acclimated to it. Good stands of crownvetch have been maintained under a two-cut hay harvest and under lenient grazing.

Festulolium (Festulolium spp.)

Festulolium grass species are hybrids derived from crossing meadow fescue and ryegrass. They are bunchgrasses suitable for hay, silage, or pasture. The meadow fescue parent contributes midsummer growth, winter hardiness, and drought tolerance while the ryegrass parent contributes rapid establishment and forage quality. Winterhardiness of festulolium is lower than meadow fescue, but generally better than that of perennial ryegrass. Festulolium is best adapted to the northern half and may be better adapted to southern regions than perennial ryegrass. It grows especially well in the spring and produces palatable forage with high nutritive value similar to that of perennial ryegrass.

Festulolium yields well under good fertility when moisture is adequate. Like perennial ryegrass, it is a vigourous establisher. Because it is less winter hardy than other grasses, festulolium is best seeded in combination with other grasses and legumes. It can be grown on occasionally wet soils. Compared with orchardgrass, it is lower yielding, less competitive with legumes, and later to mature. Like orchardgrass, festulolium can withstand frequent cutting or grazing. It is difficult to cut with a sickle bar mower and is slower to dry than other grasses, so it is better suited for grazing, greenchopping, and silage harvesting than for dry hay.

Festulolium varieties can differ markedly in winterhardiness and recovery from winter injury.

Kentucky Bluegrass (Poa pratensis L.)

Kentucky bluegrass is a long-lived perennial grass especially well suited to pastures. Its low growth habit makes it less suited to hay and silage production. It forms a dense, tough sod under favorable conditions, providing good footing for grazing animals. It reproduces by seed and rhizomes. It tolerates close or frequent grazing. Kentucky bluegrass is one of the most forgiving grasses, able to tolerate and persist under a wide range of soil conditions and mismanagement, but it will respond and have better production under good management. Kentucky bluegrass grows best under cool and moist conditions and usually becomes semi-dormant during the summer. Planting good quality seed of forage Kentucky bluegrass varieties results in a more productive pasture than the naturally volunteering bluegrass.

Orchardgrass (Dactylis glomerata L.)

Orchardgrass is a versatile perennial bunch-type grass (no rhizomes) that establishes rapidly and is suitable for hay, silage, or pasture. Orchardgrass is probably the most productive cool-season grass grown, especially under good fertility management. It has rapid regrowth, produces well under intensive cutting or grazing, and attains more summer growth than most of the other cool-season grasses. Orchardgrass tolerates drought better than several other grasses. It grows best in deep, well-drained, loamy soils. Its flooding tolerance is fair in the summer but poor in the winter. Orchardgrass is especially well suited for mixtures with tall legumes, such as alfalfa and red clover. The rapid decline in palatability and quality with maturity is a limitation with this grass. Timely harvest management is essential for obtaining good quality forage.

Improved varieties of orchardgrass, with high yield potential and improved resistance to leaf diseases, are available. Maturity is an important consideration in variety selection, and a wide range in maturity is available among newer varieties. When seeding orchardgrass-legume mixtures, select varieties that match the maturity of the legume. The later-maturing varieties are best suited for matching the maturity of alfalfa and red clover and are easier to manage for timely harvest to obtain good-quality forage. In pastures, early-maturing varieties will often produce higher yields, but grazing management must be aggressive in the spring. Performance under grazing is being evaluated in several university variety trials and should be considered when selecting varieties for grazing.

Ryegrass (Lolium species)

Perennial ryegrass (*Lolium perenne L.*) is a bunch grass suitable for hay, silage, or pasture. It is best adapted to the northern half. It has not persisted well in the southern half, because it lacks heat and drought tolerance. Perennial ryegrass produces palatable forage with high nutritive value. It has a long growing season and yields well under good fertility when moisture is adequate. It is a vigourous establisher and is often used in mixtures to establish quick ground cover. Because it is less winter hardy than other grasses, perennial ryegrass is best seeded in combination with other grasses and legumes. Perennial ryegrass can be grown on occasionally wet soils. Compared with orchardgrass, it is lower yielding, less competitive with legumes, and later to mature. Like orchardgrass, perennial ryegrass can withstand frequent cutting or grazing. It is difficult to cut with a sickle bar mower and is slower to dry than other grasses, so it is better suited to grazing, greenchopping, and silage harvesting than for dry hay.

Perennial ryegrass varieties can differ markedly in winter hardiness and recovery from winter injury. Maturity also differs widely among ryegrass varieties. Be sure to purchase endophyte-free seed of forage-type varieties; seed of many turf-type varieties is infected with endophyte, which can produce compounds harmful to livestock and cause a neurological condition known as ryegrass staggers. Forage-type varieties are either diploid or tetraploid. Tetraploid varieties have fewer, but larger, tillers and wider leaves, resulting in more open sods than diploids. Tetraploids are usually slightly higher in forage digestibility.

Hybrid ryegrass (*Lolium hybridum*) is achieved by crossing perennial and annual ryegrass. It generally has characteristics intermediate between those of perennial and annual ryegrass. For example, hybrid ryegrass has lower winter hardiness than perennial ryegrass but should persist longer than annual ryegrass.

Annual ryegrass (*Lolium multiflorum Lam.*), also known as Italian ryegrass, is generally annual or biennial in longevity, and can provide short-term high yields of high-quality forage. Italian ryegrass varieties differ markedly in maturity and growth habit. Some are true annuals that are early maturing, head out in every growth cycle, and complete their life cycle by late summer when planted in the spring. Other varieties are very late to mature and head out little at all the first year and may survive through the winter to produce forage the second year.

Reed Canarygrass (**Phalaris arundinacea L.**)

Reed canarygrass is a tall, leafy, coarse, high-yielding perennial grass tolerant of a wide range of soil and climatic conditions. It can be used for hay, silage, and pasture. It has a reputation for poor palatability and low forage quality. This reputation was warranted in the past because older varieties produced forage containing alkaloid compounds (bitter, complex, nitrogen-containing compounds). However, varieties are now available that make this species an acceptable animal feed, even for lactating dairy cows.

Reed canarygrass grows well in very poorly drained soils but is also productive on well-drained upland soils. It is winter hardy, drought tolerant (deep-rooted), resistant to leaf diseases, and persistent. It responds to high fertility and tolerates spring flooding, low pH, and frequent cutting or grazing. Reed canarygrass forms a dense sod. Limitations of this grass include slow establishment, expensive seed, and rapid decline in forage quality after heading.

Only low-alkaloid varieties (e.g., Palaton, Venture, Rival) are recommended if the crop is to be used as an animal feed. These varieties are palatable and equal in quality to other cool-season grasses when harvested at similar stages of maturity. Common reed canarygrass seed should be considered to contain high levels of alkaloids and is undesirable for animal feed.

Smooth Bromegrass (Bromus inermis Leyss.)

Smooth bromegrass is a leafy, sod-forming perennial grass best suited for hay, silage, and early spring pasture. It spreads by underground rhizomes and through seed dispersal. Smooth bromegrass is best adapted to well-drained silt-loam or clay-loam soils. It is a good companion with cool-season legumes. Smooth bromegrass matures somewhat later than orchardgrass in the spring and makes less summer growth than orchardgrass. It is very winter hardy and, because of its deep root system, will survive periods of drought.

Smooth bromegrass produces excellent quality forage, especially if harvested in the early heading stage. It is adversely affected by cutting or grazing when the stems are elongating rapidly (jointing stage) and is less tolerant of frequent cutting. It should be harvested for hay in the early heading stage for best recovery growth. Fluffy seed makes this grass difficult to drill unless mixed with a carrier. It is susceptible to leaf diseases.

Improved high-yielding and persistent varieties are available. Some varieties are more resistant to brown leaf spot. These improved varieties start growing earlier in the spring and stay green longer than common bromegrass, which has uncertain genetic makeup.

Tall Fescue (Festuca Arundinacea Schreb.)

Tall fescue is a deep-rooted, long-lived, sod-forming grass that spreads by short rhizomes. It is suitable for hay, silage, or pasture. Tall fescue is the best cool-season grass for stockpiled pasture or field-stored hay for winter feeding. It is widely adapted and persists on acidic, wet soils of shale origin. Tall fescue is drought resistant and survives under low fertility conditions and abusive management. It is ideal for waterways, ditch and pond banks, farm lots, and lanes. It is the best grass for areas of heavy livestock and machinery traffic.

Most of the tall fescue in older permanent pastures contains a fungus (endophyte) growing inside the plant. The fungal endophyte produces alkaloid compounds that reduce palatability in the summer and result in poor animal performance. Several health problems may develop in animals grazing endophyte-infected tall fescue, especially breeding animals. Deep-rooted legumes should be included with tall fescue if it is to be used in the summer. Legumes improve animal performance, increase forage production during the summer, and dilute the toxic effect of the endophyte when it is present. Newer endophyte-free varieties or varieties with very low endophyte levels (less than 5per cent) are recommended if stands are to be used for animal feed. In addition, varieties are available with non-toxic endophytes that do not produce harmful alkaloids. Kentucky-31 is the most widely grown variety, but most seed sources of this variety are highly infected with the toxic endophyte fungus and should not be planted for livestock feed. When buying seed, make sure the tag states that the seed is endophyte-free or has a very low percentage of infected seed, or contains non-toxic endophyte only. Because endophyte-free varieties are less stress tolerant than endophyte-infected varieties, they should be managed more carefully.

Timothy (Phleum Pratense L.)

Timothy is a hardy perennial bunchgrass that grows best in cool climates. It generally grows better in northern. Its shallow root system makes it unsuitable for droughty soils. It produces most of its annual yield in the first crop. Summer regrowth is often limited because of timothy's intolerance to hot and dry conditions. Timothy is used primarily for hay and is especially popular for horses. It requires fairly well-drained soils. Timothy is less competitive with legumes than most other cool-season grasses. It is adversely affected by cutting or grazing when the stems are elongating rapidly (jointing stage) and is less tolerant of frequent cutting. It should be harvested for hay in the early heading stage for best recovery growth.

Chicory (Cichorium Intybus L.)

It is a perennial forb that is best utilized for rotational grazing. Chicory requires well-drained to moderately drained soils, medium to high fertility, and a pH of 5.5 or greater. It produces leafy growth with higher nutritive value and higher mineral content than alfalfa or cool-season grasses. Its deep taproot provides drought tolerance and good growth from spring through the summer slump period of cool-season grasses. Animal performance on forage chicory has been exceptional. An important management consideration is to prevent bolting of stems in the spring. Stubble heights greater than 1.5 inches or a rest period longer than 25 days can allow stems to bolt (a period of rapid stem growth). The thick taproot of chicory can be exposed and damaged by overgrazing, excessive hoof traffic, and frost heaving. Chicory requires nitrogen fertilization rates similar to those used for cool-season grasses.

Sweetclover (Melilotus Species)

The cultivated sweetclovers, both yellow (*Melilotus officinalis Lam.*) and white-flowered types (*Melilotus alba Medik.*), are not perennial species, but typically biennial. They are excellent for soil improvement and as nectar-producing plants for honey bees. Sweetclover is not generally grown for forage but has been used for pasture and even hay or silage. The plant contains coumarin, which reduces palatability to livestock. Coumarin also can be converted to dicoumarol during heating and spoilage of hay or silage, which reduces blood-clotting ability and can result in livestock death. Low coumarin varieties have been developed. Sweetclover requires well-drained soils with near neutral pH (6.5 or higher). It is drought tolerant and winterhardy.

PRE-ESTABLISHMENT FERTILIZATION AND LIMING

Soil Ph

Proper soil pH and fertility are essential for optimum economic forage production. Take a soil test to determine soil pH and nutrient status at least six months before seeding. This allows time to correct deficiencies in the topsoil zone. The topsoil in fields with acidic subsoils should be maintained at higher pH than the topsoil in fields with neutral or alkaline subsoils.

Topsoil pH Levels for Forages

- pH 6.8 for alfalfa on mineral soils with subsoil pH less than 6.0
- pH 6.5 for other forage legumes and grasses on mineral soils with subsoil pH less than 6.0
- pH 6.5 for alfalfa on mineral soils with subsoil pH greater than 6.0
- pH 6.0 for other forage legumes and grasses on mineral soils with subsoil pH greater than 6.0

Soil pH should be corrected by application of lime when topsoil pH falls 0.2 to 0.3 pH units below the recommended levels. With conventional tillage plantings, soil samples should be taken to an eight-inch depth, and lime should be incorporated and mixed well in the soii at least six months before seeding. If more than four tons per acre of lime are required, half the amount should be incorporated deeply and the other half incorporated lightly into the top two inches. If low rates of lime are recommended or if a split application is not possible, the lime should be worked into the surface rather than plowed down. This assures a proper pH in the surface soil where seedling roots develop and where nodulation begins in legumes.

Phosphorus and Potassium

Corrective applications of phosphorus and potassium should be applied prior to seeding, regardless of the seeding method used; however, fertilizer applications incorporated ahead of seeding are more efficient than similar rates not incorporated. This is especially true for phosphorus. Phosphorus and potassium fertilizer recommendations for forages are provided.

Pre-Establishment Fertilization for No-Till

For no-till seedings, take soil samples to a four-inch depth to determine pH and lime needs, and to a normal eight-inch depth to determine phosphorus and potassium needs. If possible, make corrective applications of lime, phosphorus, and potassium earlier in the crop rotation when tillage can be used to incorporate and thoroughly mix these nutrients throughout the soil. When this is not feasible, be sure to make lime, phosphorus, and potassium applications at least eight months or more ahead of seeding to obtain the desired soil test levels in the upper rooting zone. Use the finest grade of lime available at a reasonable price when surface applications are made.

Lime and phosphorus move slowly through the soil profile. Once soil pH, phosphorus, and potassium are at optimum levels, surface applications of lime and fertilizers maintain these levels. Attempts to establish productive forages often fail where pH, phosphorus, or potassium soil-test values are below recommended levels, even when corrective applications of these nutrients are surface applied or partially incorporated just before seeding.

Starter Nitrogen

Seedling vigour of cool-season forage grasses is enhanced on many soils by nitrogen applied at seeding time. Apply nitrogen at 10 to 20 lbs per acre when seeding grass-legume mixtures, and 30 lbs per acre when seeding pure grass stands. Starter nitrogen applications of 10 lbs per acre may be beneficial with pure legume seedings, especially under cool conditions and on soils low in nitrogen. Manure applications incorporated ahead of seeding can also be beneficial to seedling establishment of forages, including alfalfa. Obtain a manure nutrient analysis and base application rates on soil-test results.

Stand Establishment

Establishing a good stand is critical for profitable forage production and requires attention to details for success. As discussed previously, begin by selecting species adapted to soils where they will be grown. Plan well ahead of time so corrective lime applications have time to neutralize soil acidity, and soil fertility deficiencies can be corrected. Make sure fields are free of any herbicide carryover that can harm forage seedlings.

About 20 to 25 seedling plants per square foot in the seeding year usually results in good stands the following year. Once established, a stand having about six grass and/or legume plants per square foot is generally adequate for good yields. The guidelines presented here greatly improve the likelihood of successful establishment of productive forage stands.

Crop Rotation and Autotoxicity

Crop rotation is an important management tool for improving forage productivity, especially when seeding forage legumes. Crop rotation reduces disease and insect problems. Seeding alfalfa after alfalfa is especially risky because old stands of alfalfa release a toxin that reduces germination and growth of new alfalfa seedlings. This is especially true on heavy-textured soils. Disease pathogens accumulate and can cause stand establishment failures when seeding into a field that was not rotated out of alfalfa. Rotating to another crop for at least one year before re-establishing a new alfalfa stand is the best practice. If that is not possible, chemically kill the old alfalfa in the fall and seed the next spring, or spring kill and seed in late summer.

Seed Quality

High-quality seed of adapted species and varieties should be used. Seed lots should be free of weed seed and other crop seed, and contain only minimal amounts of inert matter. Certified seed is the best assurance of securing high-quality seed of the variety of choice. Purchased seed accounts for just 20per cent or less of the total cost of stand establishment. Buying cheap seed and seed of older varieties is a false and short-lived economy. Always compare seed price on the basis of cost per pound of pure live seed, calculated as follows:

- percent purity = 100per cent – (percent inert matter + percent other crop seed + percent weed seed)
- percent pure live seed (PLS) = percent germination x percent purity
- lbs of PLS = lbs of bulk seed × percent PLS

Seed Inoculation

Legume seed must be inoculated with the proper nitrogen-fixing bacteria prior to seeding to assure good nodulation. Inoculation is especially important when seeding legumes in soils where they have not been grown for several years. Because not all legume species are colonized by the same strains of nitrogen-fixing bacteria, be sure to purchase the proper type of inoculum for the forage legume to be planted. Verify the inoculant expiration date and make sure it was stored in a cool, dry place. Because many seed suppliers distribute pre-inoculated seed, check the expiration date and reinoculate if necessary. If in doubt, reinoculate the seed before planting. The seed should be slightly damp and sticky before adding the inoculant. This can be accomplished with a syrup/ water mixture or a commercial sticker solution. Soft drinks are also effective as sticking agents. Protect inoculants and inoculated seed from sun and heat as much as possible and plant soon after inoculation.

Seed Treatments

Fungicide-treated seed is highly recommended for alfalfa and may be useful for red clover. Apron (metalaxyl) and Apron XL (mefanoxam) are systemic fungicides for controlling seedling damping-off diseases caused by *Pythium* and *Phytophthora* during the first four weeks after seeding. These pathogens kill legume seedlings and cause establishment problems in wet soils. Many companies are marketing alfalfa seed treated with either Apron or Apron XL. Various other seed treatments and coatings are sometimes added to forage seed. Always consider the cost vs. the opportunities for benefit.

It is very important to calibrate seeders appropriately. For example, lime coatings can account for up to one-third of the weight of the seed, so the actual number of seeds planted can be drastically affected when seeding on a weight basis. In addition, some seed coatings affect the flowability of seed, which can dramatically affect the seeding rate output of a planter. The manufacturers' seeding calibrations are likely to not hold true for coated seed.

Spring Seedings

In most years, spring seeding should be completed by early May in northern rergion and by late April in southern region. With early seeding, the plants become well established before the warm and dry summer months. Weed pressure increases with delayed seeding. Annual grassy weeds can be especially troublesome with delayed spring seedings. Herbicides are usually essential when seeding late in the spring.

Direct seeding without a companion crop in the spring allows growers to harvest two or three crops of high-quality forage in the seeding year, particularly when seeding alfalfa and red clover. Select fields with little erosion potential when direct seeding into a tilled seedbed. Weed control is important during early establishment when direct seeding pure legume stands. Several preplant and post-emergent herbicide options are available for pure legume seedings.

Small grain companion crop seedings are successful when managed properly. Companion crops reduce erosion in conventional seedings and help minimize weed competition. Companion crops usually increase total forage tonnage in the seeding year, but forage quality will be lower than direct seeded legumes. When seeding with a small grain companion crop, take precautions to reduce excessive competition, which may lead to establishment failures:

- Spring oats and triticale are the least competitive, while fall-planted winter cereals are often too competitive.
- Use early-maturing, short, and stiff-strawed small grain varieties.
- Sow companion small grains at 1.5 to 2.0 bushels per acre.
- Remove small grain companions as early pasture or silage.
- Do not apply additional nitrogen for the companion crop.

Where the need for erosion control suggests use of a companion crop, but high-quality legume forage is desired the first year, seed oats as a companion and kill it at four to eight inches with a post-emergent grass herbicide. The oats will suppress early weed growth, provide erosion protection, and protect seedlings from wind damage. After oats are killed, the legume forage will perform about the same as in a direct seeding.

Late Summer Seedings

Late summer is an excellent time to establish many forage species, provided sufficient soil moisture is available. August is the preferred time for late summer seeding because it allows enough time for plant establishment before winter. Do not use a companion crop with August seedings because it will compete for soil moisture and can slow forage seedling growth to the point where the stand will not become established well enough to survive the winter.

Sclerotinia crown and stem rot is a serious disease threat when seeding alfalfa and clovers in late summer. The risk of infection is greatest in fields where forage legumes have been grown recently and minimum tillage is used. Sclerotinia infects seedlings in the fall, but injury is not visible until plants begin to die in late winter and early spring. Crop rotation, conventional tillage plantings, and seeding by early August reduce the risk of severe damage from this disease. A limited number of alfalfa varieties have some resistance to this disease.

Conventional Tillage Seeding

The ideal seedbed for conventional seedings is smooth, firm, and weed-free. Don't overwork the soil. Too much tillage depletes moisture and increases the risk of surface crusting. Firm the seedbed before seeding to ensure good seed-soil contact and reduce the rate of drying in the seed zone. Cultipackers and cultimulchers are excellent implements for firming the soil. The lack of a firm seedbed is a major cause of establishment failures. The soil should be firm enough at planting so that a footprint is no deeper than ½ to ¾ inch.

Seeding depth for most cool-season forages is ¼ to ½ inch on clay and loam soils. On sandy soils, seed can be placed ½ to ¾ inch deep. Seeding too deep is one of the most common reasons for seeding failures.

Seeding equipment. Forage stands can be established with many different types of drills and seeders, provided they are adjusted to plant seed at an accurate depth and in firm contact with the soil. When seeding into a tilled seedbed, drills with press wheels are an excellent choice. If the seeder is not equipped with press wheels, cultipack before and after seeding in the same direction as the seeder was driven. This assures that seed is covered and in firm contact with the soil. Cultipacker seeders, such as the Brillion seeder, provide accurate and consistent seed placement in tilled seedbeds.

Fluid seeding can be used to seed forage legumes. Seed is distributed in a carrier of water or in a fertilizer solution. Custom application is recommended because it requires special equipment for good seed suspension and distribution. Prepare a firm seedbed and cultipack after the seed is "sprayed" on. For fluid seeding, seed should be mixed into solution at the field and applied immediately. Some producers are also having success with seeding legumes through dry fertilizer air spreaders, with cultipacking before and after the seed is broadcast.

No-Till and Minimum-Till Seeding

Many producers are successfully adopting minimum and no-tillage practices for establishing forage crops. Advantages include soil conservation, reduced moisture loss, lower fuel and labor requirements, and seeding on a firm seedbed. All forage species can be seeded no-till. Species such as red clover, which have good seedling vigour, are the easiest to establish. No-till forage seedings are most successful on silt loam soils with good drainage. Consistent results are more difficult on clay soils or poorly drained soils. Weed control and sod suppression is essential for successful no-till establishment, because most forage crops are not competitive in the seedling stage.

Seed placement is critical with reduced tillage. It is very easy to plant seeds too deep with no-till drills. A relatively level seedbed improves seed placement. A light disking may be necessary before attempting to seed. Plant seed shallow (¼ to ½ inch, in most cases) in firm contact with the soil. Crop residue must be managed to obtain good seed-soil contact. Chisel plowing or disking usually chops residue finely enough for conventional drills to be effective. When residue levels are greater than 35per cent, no-till drills are recommended.

For no-till planting following corn, plant as soon as the soil surface is dry enough for good soil flow around the drill openers and good closure of the furrow. Perennial weeds should be eradicated in the previous corn crop. If perennial weeds are still present, apply glyphosate before seeding. If any grassy weeds or winter annual broadleaf weeds are present in the field, use paraquat or glyphosate before seeding. Most drills can handle corn grain residue, but removal of some of the residue (e.g., for bedding) often increases the uniformity of stand establishment. Most drills do not perform as well when corn stalks are chopped and left on the soil surface. Be sure to avoid problems with carryover of triazine residue from the previous corn crop.

No-till planting following small grains is a good way to conserve valuable moisture. Weeds should be effectively controlled in the small grain crop. Ideally, wait to plant the forage crop until at least ½ inch of rain has fallen postharvest to stimulate germination of volunteer small-grain seeds and weeds; however, do not delay planting beyond the recommended seeding date for your area. Burn down any weeds and volunteer small grain seedlings before seeding the forage crop. Glyphosate can be used if thistles, Johnsongrass, or other perennial or biennial weeds are present in the small grain stubble. Remove straw after small grain harvest. It is not necessary to clip and remove stubble; however, it may be removed if additional straw is desired. Do not clip stubble and leave it in the field, as it may interfere with seed-soil contact when seeding forages. If volunteer small grains become a problem after seeding, apply a selective grass herbicide to pure legume seedings to remove excessive competition.

Insect control can be a serious problem in no-till seedings, especially those seeded into old sods. Slugs can be especially troublesome where excessive residue is present from heavy rates of manure applied in previous years. Chemical control measures for slugs are limited to a methaldehyde bait (Deadline bullets). Lorsban-4E insecticide is registered for use during alfalfa establishment for suppression of cutworms, wireworms, and grubs.

Seeding-Year Harvest Management

Harvest management of cool-season forages during the seeding year depends on time and method of seeding, species, fertility, weather conditions, and other factors. Forages seeded in August or early September should not be harvested or clipped until the following year. For spring seedings, it is best to mechanically harvest the first growth. This is especially true for tall-growing legumes. If stands are grazed, stock fields with enough livestock to consume the available forage in less than seven days. Grazing for a longer period increases the risk of stand loss. Soils should be firm to avoid trampling damage. General harvest management guidelines for spring seedings, according to species, are presented here.

Alfalfa

Generally two harvests are possible in the seeding year when alfalfa is seeded without a companion crop; three harvests are possible with early planting and good growing conditions. The first cutting can be made 60 to 70 days after emergence. Subsequent cuttings should be made in early bloom stage, with the last harvest taken by the first week of September. Fall cutting is not advisable; even a late dormant cutting is not recommended because it increases the risk of winter heaving. When seeding with a small grain companion crop, the first harvest should be taken during the late boot or early-heading stage of the companion crop.

Birdsfoot Trefoil

Seedling growth of trefoil is much slower than alfalfa or red clover. Seeding year harvests should be delayed until the trefoil is in full bloom. Do not harvest after September 1. When seeded with a companion crop, an additional harvest after removal of the small grain is generally not advisable.

Red Clover

When seeded without a companion crop, red clover can usually be harvested twice in the year of establishment. Under good conditions, up to three harvests are possible. Harvest red clover before full bloom in the seeding year. If allowed to reach full bloom in the year of seeding, red clover often has reduced stands and yields the following year. Complete the last harvest by the first week of September.

Cool-Season Grasses Harvest management depends greatly on stand vigour and weather conditions. Most grasses establish slowly compared with alfalfa. Clipping may be necessary to prevent annual weeds from going to seed. Early clipping at 6 to 8 inches of grass growth will also promote tillering of the seedlings.

Fertilizing Established Stands

A current soil test is the best guide for a sound fertilization programme. Forages are very responsive to good fertility. Adequate levels of phosphorus and potassium are important for high productivity and persistence of legumes, especially alfalfa. Forage fertilization should be based on soil-test levels and realistic yield goals. Under hayland management, forages usually need to be topdressed annually to maintain soil nutrient levels and achieve top production potential.

Each ton of tall grass or legume forage removes approximately 13 pounds of P_2O_5 and 50 pounds of K_2O. These nutrients need to be replaced, preferably in the ratio of one part phosphate to four parts potassium. Phosphorus and potassium recommendations for forages are given in Tables 7-5 to 7-7.

Timing Topdress Applications

The timing of phosphorus and potassium applications is not critical when soil-test levels are optimum. Avoid applications with heavy equipment when the soil is not firm. Soil conditions are frequently most conducive to fertilizer applications immediately following the first cutting or in late summer and early fall. Split applications, one-half after the first cutting and one-half in late summer or early fall, may result in more efficient use of fertilizer nutrients when high rates of fertilizer are recommended. If soil-test levels are marginal to low, fall fertilization is especially important to provide nutrients such as potassium that are important for winter survival.

Nitrogen Fertilization

Where grasses are the sole or predominant forage, nitrogen fertilization is extremely important for good production. Economic returns are usually obtained with 150 to 175 pounds of nitrogen per acre per year split three times during the year—70 to 80 pounds per acre in early spring when grasses first green up and 50 pounds per acre after each cutting except after the last cutting of the season.

Legumes fix atmospheric nitrogen. Where the forage stand is more than 35per cent legumes, nitrogen should not be applied. In pastures, nitrogen application can be used strategically, so increased forage production occurs only when it will be most needed.

Table: Examples of Nitrogen Rates Recommended for Perennial Cool-Season Grass Forages.

Crop, percent legume	Yield Potential, ton/acre		
	4	6	8
	Annual Application (lb N per acre[1])		
Tall grass, less than 20per cent legume	100	140	180
Mixed tall grass-legume, 20-35per cent legume	50	90	130
Mixed tall grass-legume, greater than 35per cent legume	0	0	0

Make split applications of N in the early spring and after first harvest. Liquid N should be applied in early spring or immediately following forage removal.

When applying nitrogen in the summer, keep in mind that some forms are subject to surface volatilization, resulting in loss of available nitrogen. Ammonium nitrate is the best source because surface volatilization losses are minimized; however, this formulation is now difficult to purchase.

Micronutrients

Micronutrient deficiencies are rare in most mineral soils. Micronutrient fertilization should be based on demonstrated need through soil testing and/or tissue testing. Boron may be needed when alfalfa and clover are grown on sandy soils and highly weathered soils low in organic matter. If the soil test is one part per million (ppm) or less of boron (B), or a plant tissue test shows 30 ppm or less B, then apply a fertilizer containing two pounds of B per acre. Sulfur may be needed when alfalfa and clover are grown on low organic matter soils and coarse soils when yield levels are high. Use a sulfate form of sulfur if the application is made in the spring, and an elemental form of sulfur for fall applications.

Grass Tetany

Grass tetany occurs in animals when their demands for magnesium exceed the supply. It most often occurs in the spring when high-producing animals are consuming primarily grass forage. High soil potassium tends to reduce uptake of magnesium by plants. The risk of grass tetany is reduced by not applying potash in early spring to grasses, because grasses take up more potassium than needed for growth (luxury consumption), and potassium uptake is favored by plentiful soil moisture. After the first harvest, apply needed fertilizer to maintain a balanced soil-fertility programme. It may also be helpful to feed livestock a high-magnesium supplement during spring.

Established Stand Harvest Management

Harvest management is an important tool in achieving high-quality forage, high yields, and stand persistence. Harvest management can also be used to reduce the impact of weeds, insects, and disease pests. Harvest timing is a compromise between forage yield, quality, and persistence. While forage quality decreases with maturity, dry matter yield usually increases up to full-flower stage in legumes and full-heading stage in grasses. Cutting more frequently at earlier stages of maturity results in forage with higher nutritive value but lower yield compared with cutting less frequently at more mature stages of growth.

A good compromise between forage yield, quality, and stand persistence is to harvest legumes in late-bud to early-bloom stage, and grasses in late-boot to early-heading stage. Harvesting at this stage results in the highest yields of digestible dry matter per acre. Cutting management of grass-legume mixtures should be based on the best harvest schedule for the legume.

First Harvest Timing

Make a timely first harvest to achieve the best quality possible in what is usually the largest crop of the year. Forage quality declines more rapidly with advancing maturity in the spring than it does later in the summer. Timing of the first harvest should be based on the calendar rather than on stage of maturity. Bud development and flowering are not reliable guides for proper timing of first cutting. In some years, little or no bloom is present in the spring; in others, bloom is abundant.

The first cutting of legume-grass hay meadows. Harvesting during these periods maximizes yields of digestible dry matter per acre. By using various grasses and legumes that differ in maturity development, producers can spread the optimum first cutting date over one week to 10 days.

Most grasses should be harvested in the boot stage for best forage quality; however, timothy and smooth bromegrass should not be cut until the grass is in the early heading stage. Earlier harvesting of these species may reduce regrowth and cause stand loss, because the basal buds for regrowth are not fully developed until early heading.

1. *Cutting Schedule A*: Forage cut during these periods is of high quality. Dry matter yields are lower than would be received from later harvests; however, yields of digestible dry matter per acre equal or exceed those from later harvests. Current alfalfa varieties are adapted to earlier harvest.
2. *Cutting Schedule B*: Harvesting at these dates produces medium quality forage. Digestibility is lower than from earlier harvests. These dates may be followed in these situations: For long-lay sods where it is important to keep legume stands for several years; where soil pH and fertility levels are less than optimum; where a late fall cutting may have been taken; winter injured fields; north facing slopes.

Summer Harvest Timing

Stage of growth is usually a reliable guide for timing summer harvests of legumes. Generally, summer cuttings are permitted to reach early bloom for alfalfa (approximately 35 days between cuttings) and half bloom for birdsfoot trefoil and red clover. High yields of good-quality forage can be harvested if three or more cuttings are made on a 35- to 40-day schedule. Four cuttings of alfalfa can be made on soils with good fertility without any detrimental effects on the stand. Harvest schedules for legume-grass mixtures should follow closely to what favors the legume component. Smooth bromegrass and timothy are more compatible with less intensively managed stands (three-cut schedule), while orchardgrass, perennial ryegrass, tall fescue, and reed canarygrass are adaptable to more frequent harvesting.

Intensive Cutting for High Quality

More intensive frequent cutting schedules are desirable where high forage quality is important. Shorter harvest intervals will usually shorten stand life. Allowing legume stands to reach early flower stage once during the season improves stand persistence. This can usually be achieved in late summer without great reductions in forage quality (forage fibre levels increase at a slower rate in late summer than in spring and early summer). Cutting intervals that are consistently shorter than 30 days stress legume stands because the plants do not fully replenish depleted energy reserves in the taproots and crowns. Fibre levels may be undesirably low when legumes are cut extremely early.

Some grass species can be harvested very intensively to achieve dairy-quality forage. Pure stands of orchardgrass and perennial ryegrass can be maintained on harvest intervals of 24 to 28 days under good fertility management.

Fall Harvesting

Producers often want to harvest the fall growth from forage stands, but fall harvesting usually increases the risk of legume heaving and winter kill, and interferes with accumulation of root reserves required for winter survival and growth the following spring. The need for the forage or its value should be weighed against the increased risk of stand damage.

Minimizing Fall Harvesting Hazard to Tall Legumes

- Do not harvest during late September and October. Forages are actively storing reserve carbohydrates in the crowns and roots during this period.
- If a late fall harvest is made, it should be delayed until after a killing frost. A word of caution removing topgrowth at this time can dramatically increase the risk of legume frost heaving on heavy soils. Mulching with up to four tons per acre of straw-manure or two tons per acre of old hay or straw should reduce frost-heaving potential after a late harvest.

- Late fall harvesting should only be attempted on healthy, established stands grown on well-drained soils with optimum pH and high fertility.
- Avoid fall harvesting of new seedings.
- If a mid-fall harvest is made, select fields that are well drained, have optimum pH and fertility, are planted to improved varieties having multiple pest resistance, and where at least 45 days of regrowth was allowed prior to the fall harvest.

Insect Pest Management

Management of forage insect pests is important to achieve high yields of high-quality forage. The primary insect problems are the alfalfa weevil and the potato leafhopper in alfalfa. The alfalfa weevil is primarily active in the spring. The potato leafhopper is active during the summer months and can cause severe yield and quality losses in alfalfa. New alfalfa seedings are especially vulnerable to potato leafhopper damage.

When pest populations reach or exceed action thresholds, it is economically justifiable to either harvest the crop, provided it is near the harvestable stage, or chemically treat the stand to control the pest in question. Producers should scout fields and determine if the action threshold has been exceeded.

Warm-Season Forages

Annual Lespedeza (Kummerowia *spp.)*

Annual lespedezas are spring-sown, warm-season legumes adapted to the southern third. They can be used for hay, pasture, and soil erosion control. These species are relatively low yielding, but produce non-bloating forage of high nutritive value for late summer grazing. They can be grown on acidic and low phosphorus soils; however, they will respond to both lime and phosphorus fertilization. Annual lespedezas grow best on well-drained soils. They are dependable reseeders and can persist in pastures if allowed to reseed each year. Annual lespedeza can be used effectively in pasture renovation to improve animal performance and late-summer forage production, especially in endophyte-infected tall fescue pastures. Seed should be inoculated with the proper Rhizobia the first time it is planted in a new area.

Perennial Warm-Season Grasses

Perennial warm-season grasses have potential to produce hay and pasture growth during the warm and dry midsummer months. These grasses initiate growth in late April or early May and produce 65per cent to 75per cent of their growth from mid-June to mid-August. Warm-season grasses complement cool-season grasses by providing forage when the cool-season grasses are less productive. Warm-season grasses produce well on soils with low moisture-holding capacity, low pH, and low phosphorous levels. However, they do best on deep, fertile, well-drained soils with good water-holding capacity.

Annual Forage Crops

Annual forage crops can be used effectively in a forage production system. These crops:

- Provide supplemental feed when perennial forages are less productive.
- Provide emergency feed when perennial crops fail.
- Serve as interim crops between grazing periods of perennial forages when long rest periods are needed.
- Extend the grazing season in the fall and early spring.

Most annual forage crops are best used for pasture or silage rather than for hay. Double-cropping combinations are feasible with these annual forage crops (for example, small grains followed by summer annual grasses or brassicas).

Small Grains for Forage

Spring oat It is commonly used as a companion crop for seeding forage legumes. It can be used for silage or spring and early summer pasture when sown early. Oats grazed or chopped early regrow and provide a second period of grazing or greenchop. Highest yields are achieved with a single harvest in early heading to milk stage. Oats can be used for hay; however, as with the winter cereals, oats are coarse, slow to dry, and often produce dusty hay. Producers have also successfully used oats for late fall grazing, by seeding it in August or following an early corn silage harvest. Oats have been successfully aerially seeded into standing corn in mid-August to provide high-protein forage as a supplement to the lower quality corn stover when grazed in late fall after corn grain harvest.

Winter barley: It is not as winter hardy as other winter cereal grains and is more sensitive to poorly drained soils. It can tolerate moderate droughts but does not produce well under moist, hot conditions. Barley provides good quality forage for grazing in the fall if seeded early, but it should not be grazed as close or as late in the fall as wheat or rye. Barley makes good quality silage, but is less desirable for hay after heading because the heads have awns or slender bristlelike appendages on the spikelets.

Winter wheat: It provides highly digestible fall and spring pasture. Winter wheat can be sown later in the fall than barley because it is more winter hardy and able to withstand wetter soils than barley. Wheat produces more tonnage than barley and is of higher quality than rye. With careful fall or early-spring grazing, it can be subsequently harvested for grain, silage, or hay. Varieties of winter wheat used for grain may also be used for forage.

Winter rye: It is the most winter hardy of the small grains and the most productive for pasture. Forage-type varieties are available that have greater fall growth and extend the grazing season in late fall. Although best production is on fertile, well-drained soils of medium or heavy texture, it is more productive than other small grains on soils with low pH and fertility, high clay or sand content, or poor drainage. Winter rye matures the earliest of the small grains, making it the most difficult of the small grains to manage for high quality forage in the spring. Palatability and quality of rye are unacceptable if allowed to mature past the boot stage.

Triticale: It is a hybrid of wheat and rye. Varieties are available for fall or spring seeding. Fall-seeded winter triticale varieties can be used for late fall and early-spring pasture, as well as for silage or hay. Under good management, triticale produces well. High animal performance is possible when it is harvested at the right stage. Winter triticale should be managed similarly to wheat, but matures about five to 10 days after wheat.

Mixtures of small grains or small grains with annual legumes can be used to achieve specific production objectives. For example, oats can be mixed with the winter grains to increase fall growth for grazing without sacrificing yield of the winter cereals the following spring. Small grain-annual legume mixtures are especially useful when harvested as silage. The seed cost of annual legumes is usually higher and should be weighed against the value of the harvested forage. Adding annual legumes, such as peas, improves forage quality and expands the harvest window for achieving good-quality forage. These mixtures do not yield as much as corn silage, but their production in the spring may fill an important niche in a forage system. Harvest timing should be based on the proper time for the small grain species in the mixture.

Establishment and Fertilization

Seed small grains for forage in the same way as for grain. When seeding small grains for fall pasture, plant in August to mid-September. If small grains are planted only for pasture use, use the seeding rates and apply nitrogen at a rate of 50 to 70 lbs per acre at planting time.

Harvest Management

For the best compromise between yield and quality, harvest oats, barley, and wheat in the early heading stage. Although harvesting later (up to early milk stage) increases tonnage, quality declines rapidly. Triticale should be harvested in the late boot to early heading stage. Rye should be harvested in the boot stage to avoid palatability problems and large reductions in forage quality. Always use a mower conditioner to increase the drying rate of small grains.

Grazing Management

Fall and spring grazing of small grains should begin when sufficient growth is available to support livestock. In the fall, graze only early-seeded small grains. Begin grazing winter cereals when six inches of growth is available and leave three inches of stubble after grazing. Heavy fall grazing can increase the risk of winterkill. Do not graze when the small winter grain is dormant or when the ground is frozen if subsequent spring growth and/or grain production is desired. In the spring, graze only when fields are firm. Heavy or late-spring grazing greatly reduces grain yields. Remove livestock from small grain fields to be harvested for grain as soon as the plants begin stem elongation.

Animal Health Concerns with Small Grains

Animal health hazards are not as common with the small grains as they are with the sorghum species grasses; however, the following precautions should be taken:

- Supplement lush spring pastures with high-magnesium mineral blocks or mineral-salt mixes to reduce the risk of grass tetany.
- When using seed treated with fungicides, observe harvest and grazing restrictions on the label.
- Remove lactating dairy animals from small grain pastures two hours before milking to reduce the problem of off-flavoured milk.
- Split nitrogen applications to avoid nitrate poisoning.

Summer-Annual Grasses

These grasses grow rapidly in late spring and summer and, when managed properly, provide high-quality forage. They are well suited as supplemental forages during hot, dry periods when perennial cool-season forages are less productive. Because the need for extra forage usually becomes apparent after row crops have been planted in early spring, summer-annual grasses are a good double-crop option when planted after a small grain crop. They have the potential to produce forage yields of three tons of dry matter per acre within 45 to 50 days. With the exception of pearl millet, the summer-annual grasses are members of the sorghum family and have the potential for prussic acid poisoning.

Sudangrass: It is fine-stemmed, leafy, and grows three to eight feet tall. Sudangrass regrows following each harvest, until cool temperatures or lack of moisture inhibit growth. It is the preferred summer-annual grass for pasture and can be used for hay. Solid stands grow shorter than when seeded in rows. Sudangrass usually contains lower levels of prussic acid and is usually lower yielding than the other sorghum family grasses.

Sudangrass hybrids are generally slightly higher yielding and have slightly higher prussic acid levels than sudangrass at comparable stages of growth.

Sorghum-sudangrass hybrids are crosses of sorghum with sudangrass. They resemble sudangrass in growth habit, but are generally taller, have larger stems and leaves, and are higher yielding. This grass can become coarse and unpalatable if not properly utilized. It is not as well suited for hay production as sudangrass. Sorghum-sudangrass hybrids regrow following each harvest, barring restrictive environmental conditions.

New varieties with higher digestibility are available, known as brown midrib varieties. Those varieties have a characteristic brown discolouration on the main vein (midrib) of the leaves, which is a marker for the mutation for lower lignin content. The brown midrib varieties have been shown to have greater animal preference and animal performance (intake and gains) compared with the normal varieties.

Forage sorghum grows six to 15 feet tall and has potential for high yields. It is utilized as a one-cut silage or greenchop crop. Forage sorghum produces silage containing more digestible energy than legume and cool-season grass silage. Making high-quality silage from forage sorghum is generally easier than from forage legumes because of the high levels of nonstructural carbohydrates, which enhance fermentation.

The high-energy, low-protein characteristics of forage sorghum silage make it a good supplement for high-protein forage legumes. Because the feeding value of forage sorghum silage is considered to be about 85per cent that of corn silage, corn silage is usually the preferred high-energy silage grown. Forage sorghum has the potential, however, to grow better than corn on light-textured, shallow soils that tend to be droughty. Brown midrib varieties with higher digestibility are available, as described earlier for the sorghum-sudangrass hybrids.

Pearl millet is not in the sorghum family, and prussic acid is not produced in the plant. It tends to have smaller stems and is leafier than the sorghum grasses. Pearl millet regrows after each harvest, but not as rapidly as sudangrass or sorghum-sudangrass hybrids. Its regrowth may also be more sensitive to cutting height than sudangrass. Other types of millets include German, Foxtail, and Japanese millet. German and Foxtail millet do not regrow after harvest. Japanese millet grows best in wet soils.

Summer-annual grass-legume mixtures are marketed by some seed dealers. The legumes (e.g., field pea, soybean) generally improve protein content compared with summer-annual grasses grown alone. The annual legumes included in these mixtures would be present in the first growth only; regrowth would occur only from the grasses. The additional cost of the legume seed should be weighed against the improved forage quality potential.

Establishment

Summer-annual grasses require well-drained to moderately well-drained soils. They grow best in warm weather and should be planted from about two weeks after corn planting until the end of June in northern region and mid-

July in southern region. Soil temperatures should be at least 60°F to 65°F. Late plantings (after mid-July) shorten the growing season and may result in low yields because of poor establishment in dry soils in the summer followed by cool fall temperatures. Making two seedings about three weeks apart staggers the maturities and makes rotational grazing or harvest timing easier to manage.

Seeds should be planted ½ to one-inch deep on a well-prepared, firm, and moist seedbed. The seed can be broadcast and harrowed, lightly disked, or seeded with a grain drill. Forage sorghums should be planted in rows with row-crop planters to facilitate harvest and minimize lodging. These summer-annual grasses may also be established in grass sods or stubble with no-till equipment, but this is less desirable than conventional seedbed preparation.

Fertilization

Determine lime and fertilizer needs by soil test. Fertilization is similar to that used to grow 100- to 150-bushels-per-acre corn. Incorporate fertilizer in the soil prior to seeding or apply at least six months before for no-till seedings. The soil pH should be maintained between 6.0 and 6.5 for best results.

Nitrogen fertilization is critical to achieve high yields and varies by previous crop. Split applications of nitrogen should be made—half applied prior to seeding, and the remainder divided equally and applied after each cutting or grazing to achieve the most efficient use. Keep in mind possible volatilization losses of some forms of nitrogen when applied in the summer.

Harvest Management

Direct-cut silage: Forage sorghum and sorghum-sudangrass hybrids are well suited as silage crops. Harvesting forage sorghum in the dough stage and sorghum-sudangrass in the heading stage should provide sufficient dry matter content for ensiling without wilting.

Greenchop or wilted silage: Sudangrass should be cut at 18 to 40 inches of growth. Sorghum-sudangrass hybrids should be cut when at least 30 inches tall, and pearl millet cut in late-boot to early-bloom stage.

Hay: Sudangrass, sorghum-sudangrass hybrids, and pearl millet can be cut for hay. Harvest when the grasses are vegetative (before heading) or the plant reaches a height of four feet. Always use a hay conditioner to mow and crush the stems to improve drying. Even with a hay conditioner, it is difficult to field cure these grasses adequately for safe storage as hay.

Grazing: All of the summer-annuals, except forage sorghum, are suitable for grazing. Sufficient height must be achieved before grazing to reduce animal health problems and to optimize production. Grazing plants that are less than 18 inches tall will weaken them, resulting in delayed regrowth. The chance of prussic acid poisoning is higher when grazing the sorghums before a full 18 inches of growth is present. Grasses in the vegetative stage are more palatable and nutritious. Trampling and wastage increases when grazing is delayed past the boot stage. Plants reach the grazeable height of 18 to 30 inches about six to eight weeks after planting. Rotational grazing or strip grazing management should be practiced. A sufficient number of animals should be placed on the pasture to graze the grass down in less than 10 days. After grazing, clip the residue to about eight inches if old stems remain. This improves forage quality for the next grazing period.

Animal Health Concerns

Prussic acid poisoning can occur when feeding sudangrass, sorghum-sudangrass hybrids, forage sorghum, or grain sorghum. These species contain varying concentrations of cyanogenic glucosides, which are converted to prussic acid, also known as hydrogen cyanide (HCN). As ruminants consume forage containing high levels of cyanide-producing compounds, prussic acid is released in the rumen and absorbed into the bloodstream, where it binds hemoglobin and interferes with oxygen transfer. The animal soon dies of asphyxiation. Prussic acid acts rapidly, frequently killing animals in minutes. Symptoms include excess salivation, difficult breathing, staggering, convulsions, and collapse. Ruminants are more susceptible than horses or swine because cud chewing and rumen bacteria help release the cyanide.

Species and Varieties differ in Prussic acid Poisoning Potential

sudangrass varieties are low to intermediate in cyanide potential; sudangrass hybrids are intermediate; sorghum-sudangrass hybrids and forage sorghums are intermediate to high; and grain sorghum is high to very high. Piper sudangrass has low prussic acid poisoning potential, and pearl millet is virtually free of cyanogenic glucosides.

Any stress condition that retards plant growth may increase prussic acid levels in plants. Hydrogen cyanide is released when leaf cells are damaged by frost, drought, bruising, cutting, trampling, crushing, or wilting. Plants growing under high nitrogen levels or in soils deficient in soil phosphorus or potassium tend to have high levels of cyanogenic glucosides. Fresh forage is generally higher in cyanide than in silage or hay because cyanide is volatile and dissipates as the forage cures.

Reducing the Risk of Prussic Acid Poisoning in Sorghum Species

When Grazing or Greenchopping

- Graze or greenchop only when grass exceeds 18 inches in height.
- Do not graze wilted plants or plants with young tillers.
- Do not graze plants during or shortly after a drought when growth has been reduced.
- Do not graze on nights when frost is likely. High

levels of the toxic compounds are produced within hours after a frost occurs.

- Do not graze after a killing frost until the plants are dry. Wait five to seven days to allow the released cyanide to dissipate.
- After a non-killing frost, do not allow grazing because the plants usually contain high concentrations of toxic compounds. Once the first frost has occurred, grazing should not begin until five to seven days after a killing frost.
- Don't allow hungry or stressed animals to graze young sorghum grass growth.
- Feeding green-chopped, frost-damaged plants has lower risk than grazing because animals have less ability to selectively graze damaged tissue; however, the forage can still be toxic, so feed with great caution. prussic acid toxicity, as do proper levels of phosphorus and potassium in the soil.

When Making Hay or Silage

- Frost-damaged annual sorghum grasses can be made into hay with little or no risk of toxicity. When plants are wilted enough to make dry hay, most of the volatile cyanide gas will have dissipated.
- Normal silage-making allows most of the cyanide to dissipate from frost-damaged annual sorghum grasses. Delay feeding of silage for six to eight weeks after ensiling.
- Silage that likely contained high cyanide levels at harvest should be analyzed for HCN content before feeding.

Nitrate poisoning can occur under conditions of high nitrogen fertilization, heavy manure applications, drought, overcast weather, or other stress conditions that retard plant growth. Under these stressful conditions, high nitrate levels accumulate in the crop. Once forage is fed, nitrate is converted to nitrite in the animal. When nitrite levels are high, the animal cannot metabolize it quickly enough, and nitrite inhibits oxygen transport in the blood. Symptoms include rapid breathing, fast and weak heartbeat, muscle tremors, staggering, and death—if corrective steps are not taken.

The same management precautions for prussic acid poisoning help prevent nitrate poisoning. Pearl millet does accumulate high nitrate levels leading to nitrate poisoning. As mentioned above, pearl millet does not accumulate prussic acid. High nitrate levels persist when forages are cut for hay, but ensiling the crop reduces nitrates by one-half. If you suspect that forage contains high nitrate levels, have it tested before feeding.

Poisoning of horses fed sudangrass, sorghum-sudangrass hybrids, and forage sorghum has been reported. The exact cause of poisoning is not known. Do not feed horses any of these summer annual grasses.

Brassica Crops

Forage brassicas are fast-growing annual crops that are highly productive and digestible. Crude protein levels range from 15per cent to 25per cent in the herbage and 8per cent to 15per cent in the roots, depending on nitrogen fertilization rate and weather conditions. The most commonly used forage brassica crops are rape, turnip, kale, and swede. They can be grazed from 80 to 150 days after seeding, depending on species. These crops offer great potential and flexibility for improving livestock-carrying capacity from August through December. Spring-seeded brassicas boost forage supply in late summer. Summer-seeded brassicas extend the grazing season in late fall and early winter.

Rape is a short-season, leafy crop whose stems and leaves are eaten by the grazing animal; rape can also be green-chopped. It has fibrous roots, and each plant produces many stems. Rape regrows after harvest and is the easiest brassica species to manage for multiple grazings. Mature rape is excellent for fattening lambs and flushing ewes. Rape yield is generally maximized with two 90-day growth periods, but some varieties yield better with one 180-day growth period, while rape hybrids yield best with 60 days of growth for the first harvest followed by 30 days for the second harvest.

Turnip is a fast-growing crop that reaches near maximum production 80 to 90 days after seeding. Roots, stems, and leaves are grazed. The relative proportion of tops and roots varies markedly with variety, crop age, and planting date. The crude protein concentration of roots (8per cent to 10per cent CP) is approximately one-half of that in turnip top growth; however, stockpiled tops are more vulnerable to weather and pest damage than roots.

Kale is a long-season, leafy brassica that produces some of the highest yields of the brassica family when it is spring-seeded. Some varieties are very cold tolerant, which allows grazing of leaves and stems into December and January most years. Stemless varieties reach about 25 inches in height, whereas narrow stem kale grows to five feet with primary stems two inches in diameter. Stemless kale (e.g., 'Premier') establishes quickly and reaches maturity in about 90 days. Narrow stem kale is slower to establish and requires 150 to 180 days to reach maximum production.

Swede is a long-season brassica that produces a large edible root-like turnip. Swede produces higher yields than turnip, but it grows more slowly and requires 150 to 180 days to reach maximum production. Swede produces a short stem when not shaded. If plants are shaded, it produces stems 30 inches tall. Swede does not regrow after harvest.

Hybrids of Chinese cabbage with rape, turnip, or swede can also be used for forage. Research information on the production and management of these hybrids is limited.

Establishment

Brassica crops germinate quickly and can be planted to provide either summer or late fall/winter grazing:

- Plant rape, turnip, and stemless kale in the spring (mid-April through May) to provide pasture in August and September.

- Plant rape and turnips in July and August to provide grazing in November and December.
- Plant swede and kale in the spring for grazing in November and December.

Brassica crops require well-drained soils with a pH between 5.3 and 6.8 for good production. Seeding rates for rape and kalę are 3.5 to 4.0 lbs per acre, while turnip and swede are 1.5 to 2.0 lbs per acre. In the spring, use the higher side of the suggested seeding rates. Plant seeds in six- to eight-inch row spacings at ¼ to ½-inch deep in a firm seedbed. Apply 50 to 75 pounds of nitrogen per acre at seeding to stimulate establishment and seedling growth. Weed competition should be controlled during brassica establishment, otherwise stand establishment failures are very likely.

On conventionally prepared seedbeds, brassica seed can be broadcast and incorporated with cultipacking. No-till seeding into grain stubble or grass sod is recommended, but weeds and sod must be suppressed for two to three weeks to allow the brassicas to establish. Apply either paraquat or glyphosate for sod suppression. Another alternative is to apply a manure slurry or liquid nitrogen solution to burn the sod back, then no-till plant the brassica seeds. Brassicas can also be seeded with rye or other small grains to provide forage growth and protect the soil after brassicas are consumed.

Fertilization

Determine lime and fertilizer needs by a soil test. Adequate phosphorus and potassium are important for optimum growth. In addition to the nitrogen applied at planting (50 to 75 lbs per acre), another 70 lbs per acre should be applied when multiple grazings are planned with rape and turnips. This second application should be made from 60 to 80 days after seeding. Nitrogen application in a chemically suppressed grass sward tends to increase the efficacy of the suppressing herbicide. This reduces the proportion of grass in the brassica-grass sward, which is not always advantageous. Avoid excessive nitrogen and potassium fertilization to prevent animal health problems.

Harvesting

Although brassicas can be harvested for greenchop, they are most often grazed. Rotational grazing or strip grazing helps reduce trampling and waste by livestock. Graze small areas of brassicas at a time to obtain efficient utilization. Rape is most easily managed for multiple grazings. Leave six to 10 inches of stubble to promote rapid regrowth of rape. When turnips are to be grazed twice, allow only the tops to be grazed during the first grazing. Turnip regrowth is initiated at the top of the root. Both rape and turnips should have sufficient regrowth for grazing within four weeks of the first grazing.

Stockpiling these crops for grazing after maturity should only be attempted when plants are healthy and free of foliar diseases. Some varieties are more suited for stockpiling because they possess better disease resistance. Do not grow brassica crops on the same site for more than two consecutive years to prevent the buildup of pathogens that limit stand productivity.

Animal Health Concerns with Brassicas

Brassica crops are high in crude protein and energy, but low in fibre. The low fibre content results in rumen action similar to when concentrates are fed. Sufficient roughage must be supplemented when feeding brassicas to ruminant animals. If grazing animals are not managed properly, health disorders, such as bloat, atypical pneumonia, nitrate poisoning, hemolytic anemia (mainly with kale), hypothyroidism, and polioencephalomalacia, may occur.

These disorders can be avoided by following two guidelines:

- Introduce animals to brassica pastures slowly and avoid abrupt changes from dry summer pasture to lush brassica pasture. Do not turn hungry animals into brassica pasture, especially if they are not adapted to brassicas.
- Only two-thirds of the animal's diet should be comprised of brassica forage. Supplement with dry

hay or allow grazing animals access to grass pastures while grazing brassicas. No-tilling brassicas into existing grass pastures helps reduce the risk of these disorders, if sufficient grass growth is available for grazing.

Chapter 9

Multiple Cropping

Multiple cropping is the establishment and harvest of a second crop in the same season that a first crop is harvested. A number of multiple cropping systems using winter cereals (wheat, barley, rye, and spelt) as a first crop and soybean, sunflower, forage seedings and rapeseed (canola) as a second crop are either commonly used or are feasible. Corn grown for silage following a first cutting of hay and soybean after winter wheat are the most widely used double crop systems.

There are two forms of multiple cropping:

1. Double cropping and
2. Relay cropping.

With double cropping the second crop is planted following the harvest of the first. Relay cropping consists of interseeding the second crop into the first crop before it is harvested. Both systems are employed, with double cropping being the preferred system south of I-40 and relay cropping to the north of I-40.

MULTIPLE CROPPING REQUIREMENTS

Multiple cropping drastically reduces the elapsed time between successive crops and therefore can greatly increase the disease pressure for both crops. Where intense multiple cropping is practiced, the beneficial effects of crop rotation are totally negated.

There are two primary requirements for profitable multiple cropping:

- There must be adequate time for the production of a second crop.
- There must be adequate water to produce two crops, whether from stored soil moisture, rainfall, or irrigation.
- Because the soybean crop is photoperiod sensitive and matures in response to day length, it is ideally suited for multiple cropping systems where planting dates for the second crop are later in the season and can be variable due to weather.

Currently, most multiple cropping systems depend solely on a combination of rainfall and stored soil moisture to supply adequate water for two crops. Irrigation can be used as a supplement for soils with a less than adequate water supplying capacity and/or inadequate rainfall. While irrigation can greatly increase the consistency of crop yields, it also increases the cost of production. In the eastern Corn Belt, the small grain crop generally removes water from only the top foot of soil and rainfall is typically greater than 3 inches per month. If the top 3 inches of soil is dry when the second crop is planted, germination is greatly slowed until the receipt of adequate rainfall. There also must be adequate surface soil moisture to enable the root system to grow into moist soil where water availability is more consistent. Because the water requirement is so large for multiple cropping, it is generally most successful on soils with large water supplying capacities that are sometimes referred to as "good corn soils." These soils are typically deep (54" to 72") with loamy textures throughout the soil profile, and have high water supplying capacities in the range of 0.15 to 0.25 inches of available water per inch of soil. They also have good internal drainage, either natural or artificial, and little restriction to root development and water movement such as zones where the soil bulk density is greater than 1.6. Depending on the weather during the growing season, a soybean crop can produce about 2.5 bushels of soybean for each inch of rainfall and water removed from the soil. A wheat crop will usually use 6 to 8 inches of rainfall and/ or soil water. The receipt of 18 inches of rain distributed somewhat evenly during May through September will usually allow the production of a 70-bushel wheat crop followed by a 30 to 40 bushel soybean crop. Soil pH throughout the rooting zone should be in the range of 5.8 to 7.2 (plow layer 6.5 to 7.0) to allow for maximum root growth and water uptake. Soil permeability should be greater than 0.6 inches per hour to allow rainfall to move into and through the soil. The soil shrink-swell potential should be low to moderate to reduce damage to the root system as the soil dries and cracks.

Fig. Relay crop soybeans between rows of wheat, ready to harvest.

Double Cropping with Forages

Legumes and grasses (alfalfa, clovers, brome, tall fescue, orchard, timothy) in mixtures or in pure stands are sometimes broadcast seeded into small grains in late winter when the ground is still frozen. In these instances the primary purpose of the small grain is weed control, although some grain or forage is sometimes harvested. These forage seedings are usually intended for either livestock pasture, hay production or for soil improvement.

A legume cover crop, such as red clover or vetch, is sometimes interseeded into winter grains to provide a source of nitrogen (N) for a corn crop the following year. Red clover is easily established, and the amount of N produced by one year of growth is sometimes adequate to support a normal corn crop. Herbicides that could injure the clover must be avoided if weeds are a problem in the small grain. Legume cover crops are also used intermittently for hay or grazing, but this would reduce their N value for a subsequent corn crop. In most situations, a legume cover crop is used to benefit the following crop, but its value as a nitrogen source varies, and the following crop may need additional N fertilizer. With emphasis on no-till crop production, the N value of various legume cover crops is reduced by 40 to 70 percent, because the legume residue is not incorporated into the soil.

Double Cropping Wheat and Soybean

Early planting of the second crop is essential for success, which requires harvest of the wheat as early as possible. Potential double crop soybean yield decreases by one bushel for each day that planting is delayed after June 20. Early wheat harvest can be accomplished by planting an early to mid-maturity wheat variety soon after the fly-free date in the fall and harvesting when the grain moisture decreases to 18per cent to 20per cent and then using air with or without supplemental heat to dry the grain. These actions combined can save several days that would normally be used to field-dry wheat to 10 to 14 percent moisture. If a grower wishes to maximize wheat production because of the high value of wheat relative to a following crop, early harvest may be less important. Using later maturing varieties and optimum levels of N fertilizer generally increases wheat yields, but delays harvest.

Below-normal temperatures in June delay wheat maturation, which may require growers to reconsider planting the second crop. For someone considering double cropping, it may be necessary to have two wheat varieties differing in maturity available in the event that wheat maturation is delayed. Soyabean varieties with maturity ratings of 3.4 to 3.9 will usually mature before the first freezing temperature if planted in June. Other than selecting a variety that matures before the first freeze, variety selection is not as important for double cropping as it is for a full season crop.

Straw remaining after grain harvest must be managed so as not to interfere with planting the second crop. Some stubble may be left to provide mulch cover. Leaving an 8 to 12-inch stubble with the combine and baling the cut straw is an efficient practice and marketing the straw adds income from winter wheat. Alternatively, the straw can be chopped and spread evenly on the field. Usually, a no-till planter or no-till drill can plant through chopped straw, if the soil is not excessively wet or dry and hard.

Soil moisture at the time of planting the second crop is critical for its success, because average rainfall in July and August often does not replace the moisture used by the second crop. The probable soil moisture deficit is less than the potential maximum deficit, because a crop does not usually transpire moisture at a maximum rate as the soil dries. In most years, moisture used by wheat in May and June is replaced by rainfall, but in dry seasons some subsoil moisture may be used, leaving an inadequate amount of water for the second crop. Soils with low available water holding capacity are not suitable for double cropping soybeans. Generally, such soils are poorly drained, somewhat poorly drained without tile, eroded, or sandy. Growers should also be aware of the water holding capacity of their soil, and rainfall in May and June when planning to double crop soybeans after wheat. An important rule of thumb to consider is: "if June is dry, don't try to double crop." Increased nitrogen application for the small grain produces more vegetation, which increases soil moisture use. Because wheat uses moisture from the upper 8 to 12 inches of soil, growers should be aware of the moisture remaining below that depth.

Because of the short growing season remaining after wheat harvest and other time constraints, double crop soybeans should be planted no-till. The surface residue associated with no-tillage planting helps reduce moisture lost by evapouration and increases rainfall infiltration. In dry years, no-tillage planting can make a difference between satisfactory and unsatisfactory seed germination resulting from the moisture saved. A goal should be to plant the second crop on the same day the first is harvested. Narrow row, no-tillage planters equipped with residue cutting coulters and double disk openers have performed well for double cropping, but modern no-till drills are excellent implements also. Because double crop soybeans do not grow very tall, they should be planted in narrow rows (7.5") and planted at high seeding rates (minimum of 250,000 seeds per acre) to obtain maximum leaf canopy and yield.

Weed control is not especially difficult with ordinary double cropping. The break between crops allows the use of a nonselective herbicide, such as Gramoxone Extra or Glyphosate, to remove established weeds and the use of Glyphosate and glyphosate tolerant soybeans can be used to control weeds that develop following soybean planting. Occasionally, broadleaf weeds, such as Canada thistle or ragweed, become established in winter wheat fields and interfere with grain harvest or with the following soybean crop. These weeds can be controlled in the wheat crop with application of 2,4-D amine; 2,4-D ester; MCPA; dicamba; or Buctril. At the wheat hard-dough stage, 2,4-D amine can be used to control some weeds that might be a problem for the following soybeans. Always read the herbicide label to ensure compliance with requirements for use.

Relay Intercropping Wheat and Soybean

Central and northern region are near the limit where ordinary double cropping of soybeans is practical because of late harvesting of wheat and early autumn frosts. For these areas, relay intercropping offers increased yield potential because the second crop is already established at the time the first crop is harvested. Improvements in winter wheat and soybean varieties, as well as equipment and crop management, make these two species especially well-suited for relay intercropping.

For the relay intercropping system, soybeans are planted into standing wheat with a no-till planter or drill beginning in early May and can continue as long as damage to the wheat can be minimized. Early June should be the latest interplanting time because of a significant decrease in soybean yield potential after that period and increased potential to damage the wheat crop. The optimum time for interplanting soybeans into the 10-15 inch space between wheat rows is late May to very early June. Growing wheat in wide rows facilitates the interplanting of soybeans. The spacing between wheat rows may vary as needed to accommodate implement wheels. When planting wheat, plugging various seeding units allows the placement of disk openers in a pattern to accommodate tractor and planter wheels. Both wheat and soybeans can be planted with the same equipment or different planting tools may be used for the two crops. Any planting tool is satisfactory, provided that 18 to 24 wheat seeds are planted per foot of row and are evenly spaced. Row arrangement depends on available equipment and whether or not the grower emphasizes production of wheat or soybeans. Wheat row spacing has little effect on wheat yield and good yields can be obtained using wheat rows spaced 15 inches apart.

Wheat and soybean yields produced using modern relay cropping techniques and systems are typically about 90 percent and 50 percent respectively of full season yields when adequate moisture is available for the relay soybean crop. Economics have generally favored using wider or fewer wheat rows, sacrificing some wheat yield in favor of increased soybean yield.

Use of early-maturing wheat varieties is not as important for relay intercropping as for double cropping, but harvesting wheat at high moisture (18 to 20 percent) allows for early release of the soybeans and decreases interference of soybeans with wheat harvesting. Test weight and quality of wheat is improved by early harvest (higher grain moisture), but the wheat must be dried for storage. It is important to understand that somewhat vigourous wheat is required in this cropping system to suppress soybean growth until the wheat is mature. Otherwise, soybeans grow too tall and are damaged during wheat harvest. If soybeans get too tall and wide row spacings are being used, shields may be added to the combine cutter bar to cover the cycle and push the beans down to prevent cutting the tops off excessively tall soybeans while harvesting wheat.

If the wheat stand and its vigour are poor in the spring, either refrain from interplanting soybeans or wait until wheat is headed before interplanting. In several research trials, the height of early May interplanted soybeans was about two-thirds that of wheat at harvest in early July, and were damaged during wheat harvest. Best results with intercropping to date have been with soybeans planted in late May to early June when wheat heads are emerging. Traffic associated with plantings after this time tends to damage wheat and reduce its yield.

Improvements in machinery design, wider equipment, and modifications to confine traffic to the barren strips in skip-row wheat plantings eliminates some crop damage. Shields to protect wheat from tires and planter units are useful additions to the equipment when wheat rows are spaced 10 to 12 inches apart. Narrow rows should be carefully managed with respect to making wheel width and equipment adjustments to avoid damage to the wheat from soybean interplanting or to soybeans from wheat harvesting. In some cases, using narrow dual wheels on combines will reduce damage to the soybean crop during wheat harvest. Damage to soybeans increases progressively with decreasing wheat row spacing. During wheat harvests straw should be chopped and evenly spread to avoid smothering the soybeans.

It is very important that the wheat not lodge prior to harvest so that it can be harvested without excessive damage to the soybeans. To prevent lodging, use stiff straw, lodging resistant varieties and limit applications of nitrogen fertilizer. Seeding rates for soybeans should be 6 to 7 seeds per foot in 15-inch rows and 4 to 5 seeds per foot of 10-inch wide rows. Seed treatment with Apron XL or Allegiance FL is encouraged to help control root-rot diseases, and the chosen varieties should have an Rps gene and good partial resistance to Phytophthora root and stem rot. The recommended seeding and nitrogen application rates for the wheat should not be exceeded to ensure that lodging of the wheat does not occur. The least amount of wheat lodging slows harvest, reduces grain quality and increases the potential of damaging the soybean plants during wheat harvest.

In northern, full season soybean varieties have performed best in this cropping system when planted at the end of May. Later maturing varieties can be used in central. Early June plantings require slightly earlier maturing varieties than late May plantings. Short season varieties,, flower too soon after planting and do not produce enough vegetative growth to provide a complete leaf canopy for maximum sunlight interception and therefore produce poor yields.

Soil phosphorous and potassium levels should be greater than 15 and 150 ppm, respectively. Twenty to 30 pounds of nitrogen should be applied at wheat planting to encourage increased fall growth, and earlier jointing and heading in the spring. Spring nitrogen applications should be made between March 1st and April 15th at the rate of one pound per bushel of yield goal.

Weed control for the soybean crop in a relay intercropping system can by accomplished by the use of Glyphosate and Glyphosate-tolerant varieties.

Alternative Wheat/Soybean Relay Intercropping System

This system uses rows spaced 15-inches apart for both the wheat and the soybeans and can be accomplished with equipment already available on the farm. With this system the soybeans are planted in early May rather than at the end of May as a means of eliminating damage to the wheat plants while planting the soybean. A full-season soybean variety is used in this system, and the seed coated with the new, patented IntelliCoat® polymer seed coating, RELAY™ CROP, is designed to delay germination until wheat head emergence in late May/early June. The IntelliCoat polymer seed coating is a new product available from Fielders Choice in Monticello, Indiana.

In early May, wheat plants are relatively small, and there is little chance of plant damage while planting the soybeans. The tractor should have narrow tires (10" to 13") so as not to run over rows of wheat plants. Both grain drills and planters are satisfactory for this system as long as the soybean row is placed midway between the wheat rows. The major difference in this system and the normal relay intercrop system is the earlier planting of seed with a special coating to delay germination. As with the normal wheat/soybean relay intercrop system, the wheat variety must not lodge.

The varieties selected for use in this system should have an erect growth habit and medium plant height. The yield potential of wheat grown using this system is 85per cent to 90per cent of non-intercropped wheat grown in 7.5-inch wide rows. The seeding rate for 15-inch-wide wheat rows should be 18 to 24 seeds per foot of row, which cuts the seed cost per acre in half and partially compensates for the reduced wheat yield.

EMERGENCY AERIAL SEEDING

Aerial seeding is primarily an emergency seeding method for fall-seeded small grains, but can be used to save time when other crops are being harvested. The time saved can sometimes offset the increased seed and seeding costs by permitting the timely harvest of other crops.

Planned aerial seeding of small grains is sometimes made into soybeans just before leaf drop so that the soybean leaves will provide some mulch cover for the seed. Sometimes, this occurs some weeks before the fly-free date. If the seed germinates before the fly-free date, there is an increased risk of infestation of Hessian fly and viral and foliar diseases. However, there is often insufficient soil moisture and/or rainfall for germination of seed due to greatly reduced seed-to-soil contact. This usually results in delayed germination and an extended germination and emergence period. Seeding rates should be increased by at least 50per cent to compensate for these adverse seeding conditions. If the seed has not germinated by the time of soybean harvest, shallow tillage to cover the seed will improve germination.

Chapter 10

Alternative Crops

BUCKWHEAT PRODUCTION

Most Americans know of buckwheat only from its use in buckwheat pancakes. Those more familiar with the crop know it to be a versatile, easy-to-grow, short-season grain crop adapted to many regions. It tolerates poor soils and is often used as a soil-improving crop, a role it served for such notables as Thomas Jefferson and George Washington on their Virginia farms.

Buckwheat production has been limited by a small market and the crop's relatively low yield. However, it has made a niche for itself because of some desirable food characteristics and its unique status as a short-season crop that can be planted later than any other summer grain crop. In most areas where it is grown, it is used as the sole crop for a field in the summer, often as a late-planted alternative after regular crops have failed.

Buckwheat is a broadleaf, herbaceous plant that flowers prolifically over a period of several weeks. The small white flower clusters quickly develop into triangular brown seeds. The brown buckwheat grains actually consist of a true seed (groat) surrounded by a thick hull. Buckwheat is sometimes referred to as a pseudocereal because the grain is used in ways similar to cereal grains such as oats, but it is not a true cereal crop due to seed and plant type. Like soybeans, buckwheat produces flowers in an indeterminate fashion, and flowering will often occur right up until harvest or frost. At peak bloom, the green leaves of the crop canopy are almost hidden under masses of white flowers. Flowers are self-sterile and must be cross fertilized by insects or wind for seed set to occur. Cool, moist conditions also aid in seed set, but many flowers will abort.

Buckwheat emerges quickly in warm soil conditions and reaches a height of 20 to 30 inches. The plant has a fairly small, shallow rooting system, and thus is not particularly drought tolerant (but it may avoid midsummer droughts if planted late). Branches form primarily in the upper canopy. Leaves are alternate and heart-shaped, usually two to three inches in length. When seeded in narrow rows, a thick crop canopy develops within a few weeks of planting.

Buckwheat flour can be light in colour if hulls (one-fourth of the grain in weight) are completely removed before grinding the groats, but often some hull fractions remain in the ground material, giving the flour a dark colour. Whole-grain buckwheat is about 11per cent to 12per cent protein, but after removing the hull, the remaining seed is about 15per cent to 17per cent protein with 3per cent fat. Buckwheat is used most frequently for soil cover as a green manure crop or smother crop on gardens or small fields. As a green manure crop, buckwheat produces only modest biomass but offers rapid growth, improves soil tilth, and makes phosphorous more available. Quick, aggressive growth accounts for its success as a smother crop for suppressing weeds, particularly in late summer. Buckwheat is popular among beekeepers. It produces a dark-coloured honey with a distinctive flavour. An acre of buckwheat can support a hive of bees producing up to 150 pounds of honey, if prevailing weather conditions are suitable for good nectar production.

With the contract price for buckwheat grain normally close to 10 cents per pound, yields of between 350 and 800 pounds per acre or better usually are needed to make a profit.

However, buckwheat, like any alternative crop, is somewhat riskier than a traditional crop. Producers should start with a modest acreage to gain experience.

Buckwheat generates only modest grain yields in comparison to many crops, in part reflecting the very limited amount of breeding that has been done with the crop in recent decades. Typical yields are 800 to 1,200 pounds per acre with good growing conditions. Buckwheat tolerates relatively poor, infertile soils better than most grains, but yields best on medium-textured, well-drained soils. It is reportedly tolerant of poorly drained soils but should be avoided on heavy or droughty soils. It tolerates acid soils down to a pH of 5. Soils prone to surface crusting may not be the ideal choice, since buckwheat can have emergence problems when crusting occurs.

Relatively few varieties of buckwheat are readily available most farm suppliers sell a type simply known as common buckwheat, genetic material that has not been maintained as a pure variety. Most buyers, especially those exporting to Japan, will specify the variety to be used as part of the production contract. Large-seeded varieties are almost always the ones desired for food use.

Since buckwheat varieties are not hybrids, harvested seed can be successfully used for replanting the next year. For buckwheat, as with most grains, it is important to prepare a firm seedbed if the field is tilled. When no-till planting, make adjustments to ensure that the soil closes over the seed furrow.

Many growers do not fertilize buckwheat due to its relatively low value and modest fertility needs. However, for optimum yields, some fertilizer may be needed. Nitrogen fertilizer may improve growth, particularly if available soil N is depleted following wheat. Low rates of N should be used, since more than 50 pounds of nitrogen per acre may lead to lodging. Buckwheat can get by without P and K on soils testing medium to high in these nutrients, but on soils testing low in P or K, application is recommended to achieve optimum yields.

In planting buckwheat, the key is to achieve a solid, even stand, which is mainly a matter of having good soil moisture and planting at an appropriate date. Buckwheat grown for purposes other than grain harvest can be planted at any date after the frost-free date in the spring. For grain harvest, it is desirable to plant relatively late in the summer, since flowering and seed set will then be more likely to occur as the days and nights begin to cool in early fall. When planted in late July or early August, buckwheat usually matures in eight to 10 weeks.

A general recommendation for seeding rate is 700,000 plants per acre, which is about 50 to 55 pounds per acre of large seed or about 40 pounds per acre of small seed. Buckwheat can compensate somewhat for a thin stand by branching more, and as a result studies show little yield response to seeding rate.

Typical seeding depths are one to two inches, depending on depth to soil moisture. With warm soils, the seedlings will emerge in four to five days. The crop must be drilled in narrow rows, 7.5-inch widths, to obtain a good canopy for shading weeds and for optimum yield. Most standard grain drills will work effectively with buckwheat.

No herbicides are currently registered for buckwheat in the United States. Growers must rely on sensible field selection, pre-plant weed control, and establishment of a uniform, dense crop canopy to shade out weeds. Tillage shortly before planting can control existing weeds and provide a good seedbed. However, care should be taken to avoid drying out the seedbed through excessive tillage. Buckwheat will reseed itself but is easily controlled by tillage or a number of broadleaf herbicides. Thus, it is not likely to present a weed problem in a following crop.

Buckwheat has few reported pests, perhaps because the crop is not extensively grown. Reported insect pests include aphids, wireworms, and Japanese beetles. Rhizoctonia root rot may occur, but other diseases are rare. Deer or other wildlife may occasionally cause localized damage. Overall, pests are unlikely to cause any significant loss in a buckwheat field.

Buckwheat should be harvested when 90per cent to 95per cent of seeds are mature. In either case, expect the stems and some leaves to still be green at harvest, since the crop has not been bred for complete dry-down like soybeans or wheat. Combine cylinder speed should be 600 to 800 rpm, and the concave should be set initially at ½ inch, then adjusted for seed size. Avoid cracking hulls, especially if the crop is intended for the food market. A discount is usually charged if grain is brought to an elevator at more than 16per cent moisture. Temperatures greater than 110°F should be avoided in drying buckwheat. Moisture of 16per cent is fine for short-term storage. For longer-term storage, grain should be no more than 12per cent to 13per cent moisture.

Desired test weights vary by company, but are typically 45 or 46 pounds per bushel (historically, buckwheat test weight was listed as 48 pounds per bushel). Large-seeded varieties will often have a test weight only in the low 40s, so producers should be prepared to take a slight discount on test weight. Smaller-seeded varieties have higher test weights but are otherwise considered undesirable for food markets.

It is important to sell the crop within a few months of harvest, because the groats will begin to darken (they are light-coloured when fresh), and this reduces the grain's appeal to certain buyers in the food market.

Canola Production

Canola refers to rapeseed low in erucic acid (less than 2per cent) and glucosinolates (less than 30 micromoles per gram of oil-free meal). Canola has the lowest percentage concentration of saturated fatty acids of eight commonly used vegetable oils. Canola is a name registered by the Western Canadian Oilseed Crushers Association. Mixtures of canola with high erucic acid rapeseed, used in industrial oils, should not be planted. Producers need to be precise about the type of rapeseed they are planting, as well as how and where it will be marketed.

Canola consists of two species of the mustard family, Brassica campestris, called Polish turnip rape, and Brassica napus, known as Argentine rape. Both fall- and spring-planted varieties are available in both types.

Canola is susceptible to Sclerotinia wilt and should not be planted in rotation with other susceptible crops if this disease becomes a problem. Canola plantings in a field should be at least four years apart. Canola is also susceptible to white rust, downy mildew, and alternaria blackspot. When direct combining, some seed are likely to shatter and produce volunteer plants the following season. Phenoxy herbicides and many commonly used corn and soybean herbicides control volunteer canola.

Canola can be grown on most soil types but is best suited to well-drained and non-crusting loam soils. Because the seed is small, canola responds to good seedbed preparation. The seedbed should be reasonably smooth, firm, and moist within one inch of the surface. Seed should be planted 3/8- to 3/4-inch deep. Alfalfa seeding equipment works well. Some broadcast plantings have died over winter because of dry soil conditions and poor seed-soil contact. Other kinds of equipment can be used; however, they should be used with a cultipacker. Five to seven pounds of seed should be used per acre, depending on seed size and soil conditions.

Fertilization practices are similar to those used in wheat production with some additional nitrogen. Nitrogen and potassium should not be placed in direct contact with the seed, but should be broadcast. Twenty to 30 pounds of nitrogen should be applied prior to planting with an additional 80 to 120 pounds applied in early spring before growth starts. Do not use a floater to apply the spring nitrogen, as many plants under the tires may be killed if they are not in the correct growth stage or weather conditions are not ideal when the fertilizer is applied.

Because canola seedlings are very sensitive to weed competition, they should be seeded in clean fields at narrow row spacings. This results in an early leaf canopy that shades or smothers weed growth. Mixtures of canola with mustard and wild garlic reduce the market value of the crop.

Canola is ripe when plants turn a straw colour and seeds become dark brown. This occurs about July 1 for winter varieties. Combine cylinder speed should be one-half to three-fourths that used for wheat. Seed moisture should be near 11per cent for direct combining. Seed moisture should be lowered to 9per cent if it is to be stored long. Circulating air should be used in stored canola since newly harvested seed may sweat. Also, a nylon screen may be needed on bin floors and other ventilation channels to prevent the small seed from dropping through bin aerated floors. Farmers should not plan to store canola through the winter.

Most available canola varieties mature at mid-wheat harvest. Longer maturity varieties are also available. Varietal identity throughout the marketing programme may become important in the future as the oil components and characteristics of present varieties are being altered genetically. These varieties will have specific, unique uses in the world oil market. Double-cropping soybean after canola is possible. Southern Indiana farmers have gotten a five- to 10-bushel yield increase following canola rather than wheat.

Grain Sorghum Production

Grain sorghum has the potential of becoming an important crop, particularly in areas where corn cannot be grown profitably because of spring flooding, droughty soil conditions, or late planting because of unfavorable weather.

Historically, grain sorghum has not been grown in because it has not been as profitable as corn. However, new grain sorghum single-cross hybrids have a yield potential similar to corn under good conditions, and a higher yield potential under certain adverse conditions. Farmers can grow these hybrids with available equipment, and the grain can replace corn in most animal rations. Good cash markets are limited.

The immature grain of these hybrids is high in tannin and quite bitter. Birds have not caused significant damage to such hybrids tests even though adjacent fields of corn were nearly destroyed. Sorghum hybrids that did not contain the genes necessary for tannin production were destroyed in bird-infested areas. Non-bird resistant varieties developed in the west and southern United States can be grown where bird pressures are not present. Some 75- to 80-day maturity varieties are available for emergency crop situations. Yield potential is closely related to maturity in that full-season hybrids produce much better yields than short-season hybrids.

Sugarbeet Production

Sugarbeet companies have developed excellent sugarbeet varieties through breeding and field trials. Present varieties have yield potentials of more than 30 tons per acre when grown on the state's best soils using the best possible cultural practices under favorable climatic conditions.

Because sugarbeet seed is small, precision in depth of seed placement is important for the best emergence. Recommended planting depths range from ¼ inch to ¾ inches, depending upon soil temperature and moisture. Soils need to be firm enough to ensure good seed to soil contact, which enhances the seed to absorb adequate moisture for germination and early growth.

Planting before the end of April is ideal; however, May plantings have been successful. Growers plant sugarbeets any time after the middle of March as soil conditions permit to complete planting before the end of April. Some years there are a few days in late March when planting can be completed. Test the soil in late summer; in the fall, incorporate recommended rates of lime, phosphorus, and potash for fine-textured, nearly level glacial till or lakebed soils.

Spring tillage on fall plowed soil should be no more than three-inches deep. This allows for band placement of row fertilizer on wet soil with sufficient dry, loose soil to cover the fertilizer band. Covering knives and press wheels ahead of the seeding units firms loose soil over the fertilizer band. A slow planting speed is important to cover the fertilizer band and place seed properly.

A good row fertilizer programme is important when planting early in cool, wet soils. Row fertilizer should be placed two inches to the side and two inches below the seed. Often, the total phosphorus requirement of a sugarbeet crop can be applied in row fertilizer. Use 30 to 40 pounds of nitrogen (N) and up to 90 pounds of phosphorus (P_2O_5) per acre, to stimulate early growth. A sugarbeet row fertilizer can contain as little as 20 pounds of potassium (K_2O) per acre. Add manganese for soils that have a pH greater than 6.5 or a history of manganese deficiency.

An optimum row width for sugarbeet production is between 18 and 24 inches. Thirty-inch row spacing may be used to match equipment used for corn production. However, narrow widths produce larger yields and more quality sugar than wider rows.

Sugarbeets have the ability to adapt to a wide range of populations, so most producers plant to stand, i.e., do not thin by hand labor or mechanical means. Desirable stands should be between 25,000 to 40,000 plants per acre at harvest. There is no advantage to plant spacing greater than six inches. Success of stand establishment may be enhanced by the following practices:

- Providing adequate field drainage.
- Planting in April.
- Using a precision planter.
- Controlling the depth of the planting accurately.
- Placing fertilizer so it does not interfere with accurate seed depth.
- Having good seed-soil contact.
- Using recommended herbicides for weed control.
- Controlling diseases.

Cultivation needs to begin soon after beet emergence. This is useful to control weeds and aerate the soil, which helps deter seedling diseases. If row guides are established during the planting operation, accuracy and speed of cultivation improves. As cultivation progresses throughout the season, soil should be moved toward the beet row. This facilitates the harvest operation. Avoid spreading excess soil on beet crowns as this might encourage disease.

Row fertilizer should be placed to the side and two inches below the seed to prevent manganese deficiency. Apply 30 to 40 pounds of nitrogen per acre in row fertilizer. This helps prevent manganese deficiency and increases early leaf development. Row nitrogen is especially important when preplant nitrogen was not applied. For maximum crop safety, diammonium phosphate (18-46-0) and urea (45-0-0) should be avoided in a sugarbeet row fertilizer. About 50 pounds per acre of phosphorus (P_2O_5), 20 pounds per acre of potassium (K_2O), 30 to 40 pounds per acre of nitrogen (N), and six pounds of manganese (Mn) should be applied in the row fertilizer. Sometimes, sidedress nitrogen is applied in late June. In some years this late application of nitrogen may reduce sugar yield per acre.

Do not apply more nitrogen on sugarbeets. Higher rates of nitrogen reduce quality without increasing tonnage. Apply nitrogen at planting if possible. If sidedressing is necessary, sidedress when beets are very small. A survey of grower practices showed those growers with the highest sugar production per acre applied most of their nitrogen ahead of planting, while those with the lowest yields sidedressed most of their nitrogen.

Apply phosphorus and potassium according to a soil test made the previous fall. Examples of phosphorus and potassium annual recommendations are included in Tables 10-6 and 10-7.

Weeds can be a serious problem in sugarbeets because they can severely reduce yield and interfere with harvesting. Until recent years, hand hoeing and cultivation were the accepted methods for controlling weeds. With the development of monogerm seed and more effective herbicides, hand labor costs for sugarbeet production decreased greatly.

In many instances, growers have good early season weed control only to have annual grasses and broadleafs come in late. For season-long control, a pre-emergence herbicide must be followed with post-emergence ones.

Sunflower Production

The confectionary type sunflower is used for human consumption (nut meats) and bird feed, while the oil type is utilized as oil for human consumption and meal in livestock feed. Sometimes the oil type is used in bird-feed mixtures. Sunflower grows well under a wide range of soil and climatic conditions.

This crop is normally planted between May 15 and July 1, depending upon the soil conditions or other management concerns. Tolerance to a wide range of low and high temperatures contributes to sunflower adaptation in different environments. Seed will germinate at 42°F, but a 50°F temperature is more satisfactory. Temperatures must be 26°F or lower for several hours to kill mature plants. Climatic conditions during seed development affect fatty acid composition of oil which determines its ultimate human use.

The sunflower plant is not highly drought tolerant; however, it has an extensive, heavily branched tap root system which permits it to extract more soil moisture than corn roots. Short periods of drought may not greatly reduce seed yield because growth can proceed at night when transpiration is low. The critical yield period occurs 20 days before and after flowering.

Sunflower plants grow well in soils ranging in texture from sand to clay. Because present yield potential and price per hundredweight (cwt.) do not compare economically with good corn and soybean yields, single crop sunflower should be considered on soils whose average soybean yield is 25 bushel or less and corn yields 100 bushel or less per acre. Properly managed on these soils, sunflower should yield 1,500 to 1,800 pounds per acre.

Seed should be planted one to two inches deep, depending on soil-moisture conditions. A sunflower may take longer to emerge than grain crops because of slow moisture movement through the seed coat. Plant 25,000 seeds per acre from May 20 to June 15 for single crop production. Apply 75 to 100 pounds nitrogen, 50 pounds phosphorus, and 50 to 75 pounds potassium per acre to produce 1,800 to 3,000 pounds of grain.

When planting sunflower as a double crop following wheat, apply 50 to 75 pounds nitrogen prior to or immediately following planting. Granular nitrogen may reach the soil surface more easily than liquid, especially if large amounts of wheat stubble or organic matter are present. Apply herbicides according to the weeds expected. If cultivation is used for weed control in single crop, it must be done before plants are one-foot tall. Only a few herbicides are labeled for weed control in sunflower. However, sunflower seedlings are strongly rooted and usually not injured by rotary hoeing or other similar implements that may be needed to supplement herbicidal weed control in single-crop plantings.

Herbicides registered for sunflower are often used in other crops and are familiar to most growers. Treflan at one to two pints per acre, Prowl at one to three pints per acre, or Sonalan at one-half to three pints per acre may be applied prior to planting and incorporated two- to three-inches deep to control most annual grasses and some annual broadleaf weeds. These chemicals are appropriate for single-crop plantings. Treflan must be incorporated within 24 hours after application for satisfactory results. Prowl should be incorporated within the first few days (no later than seven days) after application. Use the lower rate on light-coloured soils that are low in organic matter (less than 2per cent).

Poast has label directions for post-emergence control of annual and perennial grass weeds. The rate varies from 0.5 to 2.5 pints, depending on the grass species. Perennial grass control generally requires higher rates and may require two applications. Do not apply more than 2.5 pints of Poast total in one season. Do not apply within 70 days of harvest. If no-till sunflowers are to be planted, either full-season or double-crop, Gramoxone Extra at 0.5 to 2.5 pints per acre and a non-ionic surfactant should be applied to burn down any existing vegetation. Apply the tank mix after sunflowers are planted, but before they emerge.

Any conventional grain combine can be used for harvesting with the addition of a sunflower head attachment. Long gathering pans extending ahead of the cutter bar are used to salvage shattered seed. Ten seeds per square foot equals a harvest loss of 100 pounds per acre. The price of these attachments varies, depending upon the size of combine head and manufacturer. Harvesting may start when grain moisture reaches 18per cent to 20per cent. Some moisture testers will not measure sunflower moisture; however, Dickey-John and Farmi offer a special chart and adapters for their machines.

Combine cylinder speed should be as slow as possible and still thresh seed from the head (300 to 400 RPM). Concaves are usually set wide open and fan air flow reduced approximately 50per cent.

Some drying or air movement will probably be required during storage. Natural air with no added heat should be sufficient under most conditions. Grain should be 12per cent moisture for temporary storage and 9per cent for long-time storage. Harvesting at a high moisture content (18per cent to 20per cent) normally results in higher yields, less bird damage, and less shattering or dropping of heads than when seeds are harvested at a lower moisture content.

Most oil-type varieties presently available are the result of hybridization, so insects are not required for pollinization of the flowers even though they may be helpful. If harmful insects—such as head moth—become a problem, fields can be sprayed to control this insect. Hybrids permit increased oil and grain yields in addition to increased disease resistance.

Birds can be a problem if sunflower fields are planted near a flyway or roost. Scaring devices, such as gas guns and shooting, can be used to protect production fields.

A producer should have an acceptable market established prior to growing sunflower. Sunflower can produce an excellent silage that approaches corn silage nutrient-wise but has lower tonnage.

Index